POLYMERIC FOAMS SERIES *series editor S.T. Lee*

THERMOPLASTIC FOAM PROCESSING

Principles and Development

POLYMERIC FOAMS SERIES

series editor S.T. Lee

INCLUDED TITLES

Polymeric Foams: Mechanisms and Materials
S. T. Lee and N. S. Ramesh

Thermoplastic Foam Processing: Principles and Development
Richard Gendron

POLYMERIC FOAMS SERIES *series editor S.T. Lee*

THERMOPLASTIC FOAM PROCESSING

Principles and Development

Edited by

Richard Gendron

CRC Press
Taylor & Francis Group
Boca Raton London New York

CRC Press is an imprint of the
Taylor & Francis Group, an **informa** business

CRC Press
Taylor & Francis Group
6000 Broken Sound Parkway NW, Suite 300
Boca Raton, FL 33487-2742

First issued in paperback 2019

© 2005 by Taylor & Francis Group, LLC
CRC Press is an imprint of Taylor & Francis Group, an Informa business

No claim to original U.S. Government works

ISBN-13: 978-0-8493-1701-9 (hbk)
ISBN-13: 978-0-367-39370-0 (pbk)

Library of Congress Cataloging-in-Publication Data

Thermoplastic foam processing : principles and development / edited by Richard Gendron.
    p. cm. — (Polymeric foams series)
    Includes bibliographical references and index.
    ISBN 0-8493-1701-0 (alk. paper)
    1. Thermoforming. 2. Plastics—Extrusion. I. Gendron, Richard. II. Series.

TP1151.T48T44 2004
668.4'93—dc22

2004051966

Visit the CRC Press Web site at www.crcpress.com

Library of Congress Card Number 2004051966

Visit the Taylor & Francis Web site at
http://www.taylorandfrancis.com

and the CRC Press Web site at
http://www.crcpress.com

Dedication

To Ariane and Corinne, my beloved daughters,
and to all of my collaborators' children,
with the hope that the world we are building for them
is a much, much better one.

Foreword

Even though polymer foams have been around for about half a century, a body of scientific literature that could have provided a better understanding of foam processing has only become available recently. But why does one need that kind of understanding? Being part of the industrial community, I have lived, along with my colleagues, with the traditional trial and error system for years and have gotten results. Well, the problem is that the outcome resulting from such a system is uncertain. How much time, effort, and money have we spent before getting valuable results? How many profitable opportunities have we lost?

In 1993, I came to visit the National Research Council's Industrial Materials Institute (IMI). I brought my industrial experience and some expertise in process development, but more importantly, I had many questions in mind:

- What makes a blowing agent work and how does it interact with the polymer?
- How is the viscosity of the blowing agent–polymer mixture affected, and can it be measured, instead of guessing at it by using some fudge factors?
- Under what conditions do bubbles start to form during the foaming process?
- Could I use this information to improve processing equipment, and especially to improve the design of mixers and extrusion dies?

At that time, despite their long experience in polymer extrusion and compounding, foaming was not part of the IMI researchers' preoccupations, so no definitive answers were given to any of these questions. However, being truly motivated by that challenging process, the National Research Council initiated a decade of collaboration with the foam industry, which led for instance to the creation of the FoamTech technology group and to a much more valuable and unique concentration of expertise in the foam area.

Being one of the early industrial partners who introduced polymer foam processing and its problematics to this research team, I am very pleased also to introduce this book. It summarizes the acquired knowledge of a group of dedicated researchers who became devoted over the last 10 years of hard work to the study of polymer foaming, through exploratory works or in collaboration with the scientific and industrial community engaged in this field.

Foam practitioners involved in production, research, or development will not necessarily find in this book straightforward solutions to their daily problems or specific answers to their long-lasting questions. They will find, instead, practical tools, valuable facts, and a basic understanding of many aspects of the foaming process that will indubitably help them to develop innovative solutions to many practical problems.

Louis Michel Caron
Vice President – Technologies
Valotech, Inc.

Preface

On September 25, 1996, at the Industrial Materials Institute (IMI) of the National Research Council of Canada (NRC), the FoamTech technology group was officially launched, and its first meeting, regrouping people from the foam industry and research and development (R&D) specialists, was kicked off. Responding to industrial preoccupations focused on seeking replacements for banned ozone-depleting foaming agents, a multidisciplinary research team was set up to provide its industrial partners with scientific expertise and access to the experimental facilities available in the institutes involved in this program. In addition to IMI, the Institute for Research in Construction (IRC) and the Institute for Chemical Processes and Environmental Technology (ICPET) were part of the task force that was initiated.

The core activities of the R&D program were developed in accordance with the needs expressed at that time. It became rapidly apparent that new foaming formulations and better mechanical properties, as well as a fine understanding of the rheology underlying the foam extrusion process, would be the top priorities for the NRC researchers involved in the FoamTech program. Obviously, development and validation of innovative physical foaming agent and polymer combinations would require accurate information about solubility and diffusivity and a clearer understanding of the mechanisms involved during foaming (nucleation and growth), especially when they occur under the practical conditions experienced during production. In-line monitoring, performed with noninvasive techniques, was identified as a valuable complement to the data generated off-line. The research should highlight the contribution of viscoelasticity to concerns previously addressed through thermodynamic principles only. Some research activities conducted at IMI have focused on the development of characterization methods suitable to the thermoplastic foam extrusion process. Examples of research conducted over the years include the development of a novel in-line monitoring technique based on ultrasonic sensors. The technique has provided a better understanding of the dynamics of phase separation in terms of the key parameters that govern the formation of bubbles. Another example is the use of an on-line rheometer directly installed on the extrusion line, permitting the characterization of the rheological behavior of the polymer–blowing agent mixture. These essential measurements were traditionally difficult to achieve, the volatility of the blowing agents preventing the use of traditional off-line rheometers.

The relevance of the work conducted by NRC researchers, which has contributed to improving on a continuous basis their expertise in this domain, was validated by the reputation of their industrial partners. Leistritz, 3M, Dow, Owens-Corning, Sealed Air, AtoFina, and Pactiv are some of the first and longest-lasting members that supported and benefited over the years from the work accomplished in NRC laboratories.

The topics developed in this book are intimately associated with the core activities undertaken since the creation of the foam-related research program at IMI. They reflect, therefore, past issues that have been addressed by the NRC specialists at some time, and also unanswered questions that still persist despite our laborious efforts. This book does not pretend to provide all the answers to problems formulated during the last 10 years, although it may suggest interesting paths that belong to the actual research inquiry. This book was written by experts of two kinds: Dr. Jekyll supplied the rigorous academic background, while Mr. Hyde shared experience in his favorite playground, filled with large-scale processing equipment. These experimentations were addressed with practical concerns in mind, as the NRC researchers interacted closely with the field practitioners. Nevertheless, part of their preoccupation in writing the manuscript was to provide the reader with useful tools that could be relevant for foam development and, moreover, to suggest fresh and innovative ideas that would still need dedicated efforts to be explored. These ideas should provide potential work for the next 10 years...

Acknowledgments

This book would not have been possible without the earlier involvement of my dear friend and ex-colleague, Dr. Louis E. Daigneault. Louis was the spirit and the workhorse behind the birth of the thermoplastic foam activities at IMI. I owe him many pecan and sugar pies for providing the opportunity to work on so many challenging projects that contributed to the development of such remarkable abilities. I also give special thanks to Dr. S.-T. Lee, who offered us this notable opportunity to structure our minds in order to share with others what we believe should be known about some foaming mechanisms.

I would personally like to acknowledge the great team effort of all the contributors who supported each other with encouragement and positive criticism. Special thanks to the reviewers who pointed out numerous flaws and helped us to improve our manuscripts. I also thank those many NRC employees and invited students and researchers who have been involved in the numerous foam projects conducted over the last 10 to 12 years. Sincere appreciation is also due to the NRC managers, especially Drs. Michel Dumoulin, Mike Day, and Blaise Champagne, who believed in these activities in general, and in the writing of this book in particular.

Finally, I want to thank all the researchers and friends who, in the last 10 years (or more), collaborated in one way or another, from challenging discussions to collaboration in foaming experiments, with one goal in mind: trying to demystify the science hidden behind thermoplastic foams. Many thanks to Drs. Paul Handa, Chul Park, N.S. Ramesh, Vipin Kumar, Wen Wu, and all the others who should be part of this long list. Here's hoping that one day common understanding will be shared by all of us!

Last, special thanks to my wife Lucie, who had to deal with her husband spending so much time in the basement with his computer, and to my two daughters, Ariane and Corinne, who many times agreed so gently to play with their Nintendo and let their dad work on this book.

Contributors

Martin N. Bureau   Industrial Materials Institute, National Research Council of Canada, Quebec, Canada

Michel F. Champagne   Industrial Materials Institute, National Research Council of Canada, Quebec, Canada

Richard Gendron   Industrial Materials Institute, National Research Council of Canada, Quebec, Canada

A. Victoria Nawaby   Institute for Chemical Processes and Environmental Technology, National Research Council of Canada, Ontario, Canada

Jacques Tatibouët   Industrial Materials Institute, National Research Council of Canada, Quebec, Canada

Caroline Vachon   Industrial Materials Institute, National Research Council of Canada, Quebec, Canada

Zhiyi Zhang   Institute for National Measurements Standards, National Research Council of Canada, Ontario, Canada

Contents

1

Solubility and Diffusivity

A. Victoria Nawaby and Zhiyi Zhang

CONTENTS

1.1   Introduction

Plastics play a major role in our everyday lives, ranging from basic commodity goods to more sophisticated cellular structures for drug delivery systems [1,2]. Cellular structures, or foams, were originally generated by mechanical, chemical, or physical means. However, steps taken to improve process economics, a ban on ozone depleting agents (CFCs, HCFCs), and product limitations associated with the use of organic solvents as blowing agents have resulted in the use of CO_2 and N_2 gases to create morphologies with diverse and intricate applications.

Generating foams using a physical blowing agent consists of saturating the polymer at a certain pressure and temperature followed by thorough mixing and equilibrium. The mixture is then subjected to a sudden thermodynamic change (temperature increase or pressure drop), which causes the saturated gas to escape, leaving behind a cellular structure characterized by cell size and density. In this process, closed-cell (rigid) foams are formed if the cell membranes around the bubble remain intact, and open-cell (flexible) foams are formed if the membranes rupture. Further, the physical properties of polymers can contribute to the formation of high-density rigid, low-density rigid, low-density flexible, and high-density flexible foams used as load-bearing, thermal insulation, furniture, automotive seating, and energy-absorbing materials, respectively [3].

The principal objective of this chapter is to provide an overview of the fundamental measurements of solubility and diffusivity of gases in polymers and their role and importance in foaming processes. Details such as theories, experimental methods, predication models, various physical foaming agents, and a new phenomenon, the retrograde behavior of polymer gas systems, will be presented. Further details on transport processes in polymeric medium and membrane can be found in comprehensive reviews by Crank and Park [4] and by Paul and Yampol'skii [5].

1.2   Theories and Mechanisms

Fundamental thermodynamic and physical parameters are used to determine the rate of transport of gases in polymers and their applications in industrial processes. Numerous polymer applications are dependent on the transport properties of polymers. These include separation of gas mixtures, extraction of low molecular components from polymers, impregnation of polymers with chemical additives, and generation of polymer foams. In all described cases the solubility and diffusivity are affected by the nature and

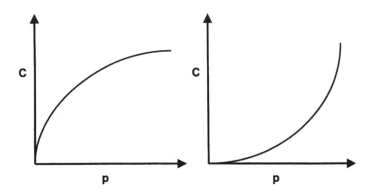

FIGURE 1.1
Non-ideal sorption behavior.

morphological aspects of the medium, such as rubbery or glassy states and crystallinity content [6].

1.2.1 Glassy and Rubbery State

The observation of gas permeation in a membrane dates back to 1829; more systematic measurements were reported in 1831 of several gases permeating into a rubber [7]. More interestingly, CO_2 was among the tested gases in the reported studies. In 1866 it was postulated that the permeation of a gas through a membrane is a combination of dissolution (sorption) and transmission (diffusion) [7]. Over the years, in addition to Fick, many contributed to the study of gas permeation in films, and in 1879 Wroblewski adapted the theories to polymer–gas systems. Sorption represents the amount of gas uptake by a polymer. From a theoretical point of view, at equilibrium and ideal sorption, the solubility coefficient (S) is a function of the concentration of gas sorbed per unit mass of polymer (C) and gas pressure (p), following Henry's law:

$$S = \frac{C}{p} \tag{1.1}$$

However, in certain polymer gas systems, sorption may not be ideal, and S (also known as the Henry's law constant) may increase or decrease with increasing C and p, resulting in solubility graphs that are concave or convex (Figure 1.1). For example, the sorption behavior of N_2 and CO_2 in polystyrene (PS) is not ideal, and polymer swelling is observed [8].

There are two well-recognized mechanisms that reduce gas solubility, pressure depression and temperature increase, which are based on Henry's law (Equation 1.1) and on the van't Hoff equation:

$$S = S_0 \exp\left(\frac{-\Delta H_s}{RT}\right) \tag{1.2}$$

S_0 is the pre-exponential factor, and ΔH_s is the heat of sorption. It should be noted that solubility of CO_2 in matrices decreases with an increase in temperature, whereas for N_2 the behavior is reversed, and an increase in solubility is observed with an increase in temperature (Figure 1.2) [9].

The transport of a gas in a polymer matrix is sensitive to polymer chain mobility, packing, and morphology [10]. Over the years there have been many reports on the solubility of gases in polymers above and below their glass transition temperature, T_g. Gas sorption in amorphous polymers at temperatures above their T_g resembles gas solubility in liquids and is adequately described by Henry's law (Equation 1.1) [11,12]. At atmospheric pressures the solubility behaves ideally; however, at very high gas pressures positive or negative deviations from ideality are observed, as represented in Figure 1.1 (e.g., interaction of silicon rubber and polycarbonate with CO_2) [13]. In glassy polymers ($T < T_g$), during gas sorption and transport two populations of gas molecules are considered to be in equilibrium with each other. The population with higher diffusional mobility follows Henry's law (Equation 1.1), and the population with restricted diffusional mobility follows the Langmuirian isotherm [14]. The Henrian contribution to gas solubility is associated with the gas molecules pushing the polymer chains apart and getting trapped between the chains. The Langmuirian contribution to solubility is due to surface absorption of the gas in the tiny holes in the

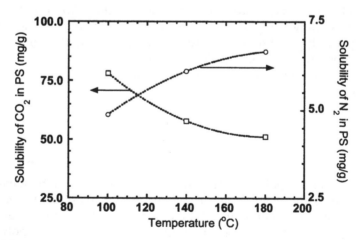

FIGURE 1.2
Solubility of carbon dioxide (CO_2) and nitrogen (N_2) in PS. (Adapted from data in Pauly, S., *Polymer Handbook*, Brandrup, J., Immergut, E.H., and Grulke, E.A., Eds., John Wiley & Sons, New York, 1999 [9].)

polymer [14]. Therefore, in glassy polymers the equilibrium gas concentration is successfully described by the dual sorption model:

$$C = C_D + C_H = Sp + \frac{C'_D bp}{1+p} \tag{1.3}$$

C_H represents the Henrian contribution to overall solubility, C_D represents the Langmuirian contribution, S is the solubility coefficient, C'_D is the Langmuir capacity constant, b is the hole affinity constant, and p is the gas pressure. Equation 1.3 was developed first by Matthes in 1944 investigating water sorption in cellulose membranes [7]. The sorption of gases in the glassy membranes or polymers has been thoroughly documented [4,7], and having information on the solubility of the gas in the polymer at various pressures and using a nonlinear regression will aid in obtaining the parameters C'_D and b simultaneously, as in the case of CO_2 sorption in polycarbonate [15].

The gas mobility in the polymer is represented by diffusion or diffusivity coefficient. The chemical potential gradient $\partial \mu$ of the penetrating gas results in the gas molecules to move within the medium of thickness l, the amount of gas transported can then be represented by Equation 1.4:

$$F = -D \frac{\partial C}{\partial x} \tag{1.4}$$

F is the rate of transfer of diffusing substance and D is the diffusion coefficient. Considering steady state conditions and maintaining a constant concentration of the penetrating gas at the surfaces $x = 0$ and $x = l$, Equation 1.4 reduces to Equation 1.5:

$$\frac{d^2 C}{dx^2} = 0 \tag{1.5}$$

Upon integration and introduction of limits, the concentration of the gas in the polymer sheet changes linearly, and thus the rate of transfer and diffusion is given by Equation 1.6:

$$F = \frac{D \left(C_i - C_f \right)}{l} \tag{1.6}$$

where C_i and C_f are the concentration of gas at the surfaces $x = 0$ and $x = l$.

Simple observation of the rate of transfer and values of the concentration at surfaces will clearly yield the diffusion coefficient of the penetrating gas across the polymer sheet. A combination of Equations 1.1 and 1.6 will yield Equation 1.7:

$$F = \frac{P(p_1 - p_2)}{l} \tag{1.7}$$

where P is the permeability of the gas to the polymer, also represented as $P = DS$, whereby $p_1 - p_2$ is the partial pressure difference across the polymer film.

1.2.2 Other Variables Affecting Sorption

Sorption and transport properties of gases in polymers are not only dependent on the state of the polymer (glassy or rubbery) but also on its morphological aspects. There are three factors which generally have the greatest influence: (1) the structure and degree of crosslinking, (2) polymer crystallinity content, and (3) plasticizers and fillers.

1.2.2.1 Structure and Degree of Crosslinking

In some cases the transport properties are affected by the polymer density, and the diffusion coefficient is reported to decrease with an increase in degree of saturation and presence of side chains where segmental mobility is reduced [7]. An increase in degree of crosslinking equally decreases the diffusion of the gas in the polymer (Figure 1.3) [9]. In cases where crosslinking affects the chemical composition of the polymer, and where the polymer is transformed into a rubbery state, a decrease in solubility is observed.

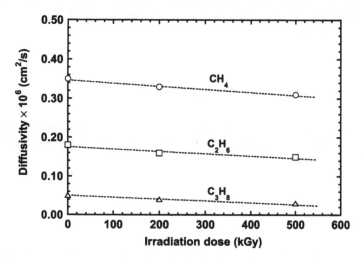

FIGURE 1.3
Diffusivity of gases in irradiation crosslinked LDPE. (Adapted from data in Pauly, S., *Polymer Handbook*, Brandrup, J., Immergut, E.H., and Grulke, E.A., Eds., John Wiley & Sons, New York, 1999 [9].)

TABLE 1.1

Solubility and Diffusivity of Various Gases in LDPE and HDPE

	Diffusivity ($\times 10^6$) (cm²/s)								
	He	O₂	Ar	CO₂	N₂	CH₄	C₂H₆	C₃H₈	SF₆
LDPE (0.914 g/cm³)	6.8	0.46	0.36	0.372	0.320	0.193	0.068	0.0322	0.0135
HDPE (0.964 g/cm³)	3.07	0.17	0.12	0.12	0.093	0.057	0.015	0.0049	0.0016
UPVC	2.8	0.012	0.0012	0.0025	0.0038	0.0013	—	—	—
PEMA	42.5	0.106	0.0208	0.0336	0.0301	—	—	—	0.00023
	Solubility ($\times 10^6$) (cm³ (273.15 K; 1.013 $\times 10^5$ Pa)/cm³ $\times$ Pa)								
	He	O₂	Ar	CO₂	N₂	CH₄	C₂H₆	C₃H₈	SF₆
LDPE (0.914 g/cm³)	0.544	0.472	0.571	2.54	0.228	1.13	7.55	21.3	0.951
HDPE (0.964 g/cm³)	0.028	0.18	1.1	0.22	0.15	0.51	3.0	8.3	0.39
UPVC	0.055	0.29	0.75	4.7	0.23	1.7	—	—	—
PEMA	0.0122	0.839	2.08	11.3	0.565	—	—	—	5.62

Source: Data from Pauly, S., *Polymer Handbook*, John Wiley & Sons, New York, 1999 [9].

Equally, if the degree of densification in the polymer decreases as a result of the crosslinking, then an increase in solubility is reported [7].

1.2.2.2 Polymer Crystallinity Content

It is now well accepted that sorption and transport of a gas in a polymer occurs in the amorphous phase of the polymer and the crystalline phase acts as a barrier for the diffusing gas [16,17]. Therefore, the solubility and diffusion coefficient in semicrystalline polymers is dependent on the amorphous fraction of the polymer. Investigation on the transport properties in semicrystalline polymers has been well documented [6,14,18–22]. For example, in semicrystalline polymers such as high-density polyethylene (HDPE) and low-density polyethylene (LDPE), the transport of gas molecules is dependent on the crystalline content of the samples (Figure 1.4 and Table 1.1) [9]. Other factors discussed in the literature are the effect of orientation and stretching, changes in free volume, and subsequent effects on the transport of gases. However, in semicrystalline polymers the observed solubilities and diffusivities are subjected to corrections with respect to the volume fractions of the amorphous phase or the crystalline phase. Generally this volume fraction is determined either by X-ray or differential scanning calorimetry (DSC) studies [20]. If φ_a is the volume fraction of the amorphous phase in the semicrystalline polymer, the solubility coefficient of the gas in the amorphous phase of the polymer is given by the following:

$$S_a = \frac{S_0}{\varphi_a} \tag{1.8}$$

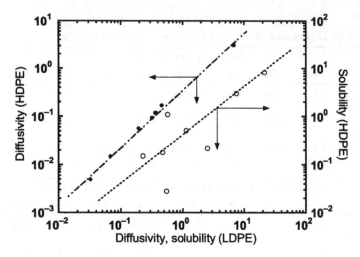

FIGURE 1.4
Solubility and diffusivity of various gases in LDPE and HDPE. (Adapted from data in Pauly, S., *Polymer Handbook*, Brandrup, J., Immergut, E.H., and Grulke, E.A., Eds., John Wiley & Sons, New York, 1999 [9].)

where S_0 is the apparent solubility (i.e., experimental data) and S_a is the solubility coefficient in the amorphous phase. Hence, the diffusion coefficients are also corrected using Equation 1.9:

$$D_a = D_0 \tau \beta \tag{1.9}$$

D_0 is the observed diffusion coefficient, and τ is a geometric tortuosity parameter related to the mean distance between crystallites, which accounts for the impedance to gas transport, a consequence of the necessity for the gas molecules to go around the crystallites in order to move through the amorphous phase (Figure 1.5). β is the chain immobilization factor, which accounts for the fact that the chains right next to the crystallites are not flexible compared to those in the amorphous phase [23]. The values of τ and β are determined using heat of solution, diffusive activation energy, and the cohesive energy density of the polymer [24]. However, in the cases where values for τ and β are not available, Equation 1.9 can be approximated as follows [23]:

$$D_a = \frac{D_0}{\varphi_a} \tag{1.10}$$

The tortuosity factor has also been reported to be proportional to the amorphous volume fraction, as represented by Equation 1.11 [7,10], where ϕ_C is the crystalline fraction in the molecule:

$$\tau = \left(1 - \varphi_C\right)^n \tag{1.11}$$

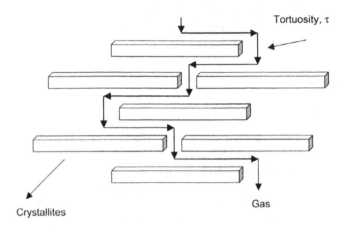

FIGURE 1.5
The tortuous path of gas in semicrystalline polymers.

For gases such as helium, oxygen, argon, carbon dioxide, nitrogen, and methane, the power of n has been reported to be close to 1; n is as high as 1.88 for some other polymer–gas systems [7].

1.2.2.3 Plasticizers and Fillers

Introduction of plasticizers will clearly influence the transport properties of the polymer, and higher diffusion coefficients are therefore observed. In the case of fillers, some are impermeable to gases, and thus a reduction in solubility and diffusivity is generally observed in these cases. There are countless papers on the effect of fillers on the transport of gases in polymers; the most recent ones have reported on the addition of titanium dioxide (TiO_2) and organically modified clay to polymer matrices, and have investigated the role of these additives in the foaming process. The solubility of CO_2 in polyethylene (PE)–titanium dioxide composites was measured, and it was observed that the apparent solubility, defined as the weight of dissolved CO_2 per unit weight of composite, decreased with an increase in TiO_2 concentration. However, the true solubility of the gas per unit weight of polymer remained constant with an increase in the TiO_2 content [25].

A review published in 2000 by Alexander and Dubois provides an excellent summary of the preparation of composite materials in addition to the uses of this new class of materials [26]. A more focused investigation on the foam processing of nanocomposites has been reported, in which the effect of the filler on the cellular structure of the polypropylene (PP)–clay nanocomposites was studied [27]. The PP nanocomposites with 2 wt% clay content foamed with CO_2 showed closed cells with pentagonal and hexagonal structures, while 7.5 wt% clay content in PP resulted in foams with spherical cells. As part of a series of ongoing studies on the effect of fillers, the effect of clay on the nanostructure control and foam processing of polycarbonate and a number of biodegradable polymers has also been investigated [28,29].

1.3   Models for Prediction

The transport and dissolution of a penetrating gas in a polymer follows the classical solution-diffusion model. Diffusion is well presented by Fick's law and sorption follows Henry's law (Equation 1.1).

Diffusion coefficients of a gas in a polymer at various temperatures can be determined either theoretically or experimentally. In the absence of data, the diffusion coefficients are in some cases estimated by theoretical means [30]. The details on the more fundamental correlations and estimation of the transport and solubility coefficients are provided in the literature [4,7]. There are generally two approaches in determining diffusion theoretically: (1) the free volume model and (2) the molecular model.

The dependence of the transport coefficient on temperature is described by Equation 1.12:

$$D = D_0 \exp\left(\frac{-E_a}{RT}\right) \qquad (1.12)$$

D_0 is the pre-exponential factor and E_a is the activation energy for diffusion. Both the free volume model and the molecular model assume the existence of microvoids or cavities. Based on the free volume model, the penetration of a gas in a polymer is regarded as a series of activated jumps from one cavity to another in the matrix, and diffusion occurs from a random fluctuation of the free volume and a redistribution of microvoids. Therefore, in any process (involving plasticizers or swelling agents), increasing the number of cavities in the polymeric matrix will result in an increase in the rate of transport, whereas crosslinking agents will decrease this rate [31]. However, the use of this model is not as attractive, due to difficulties with obtaining the physical property data for some polymers. A number of studies have demonstrated relationships between the free volume theory, the glass transition, and the degree of crosslinking of polymers, in addition to the Flory-Huggins solution theory, thus determining the diffusion coefficient of the gas [7]. The molecular model, however, assumes that if the size of the cavity is large enough, a penetrating gas when equipped with appropriate energy can jump into the neighboring void. Once a second molecule for penetration becomes available and occupies the previous cavity due to a net diffusion flux, parameters such as jump length, molecular dimensions, and activation energy of diffusion become important in this model [31]. A recent investigation has been reported on the theoretical prediction of solubility and diffusivity of ethylene in semicrystalline PE at elevated pressure and temperature using a new molecular hybrid model. Results have been validated through experimental data obtained from sorption kinetic studies [32]. The impact of average free volume element size on transport properties of CO_2 in stereoisomers of polynorborene has recently been reported and the effects of temperature investigated [33]. A constitutive equation was proposed in order to determine the viscoelastic diffusion flux. The model was subsequently used to calculate the sorption curves, and data was compared to experimental observations for a wide range of polymer solvent systems [34]. With respect to solubility, a molecular model has been proposed which predicts the solubility of gases in flexible polymers. The model is based on the ideal solubility of gases in liquids. It has been reported that the model provides a crude estimation for the solubility of the gas on the basis of only the monomer unit of the polymer and the properties of the gas. A comparison between the data obtained from the model and those obtained through experimental techniques for several gases in poly(dimethylsiloxane) (PDMS) has been reported [35]. This model predicts decreasing solubility with an increase in temperature, in addition to smaller molecules exhibiting lower solubility in polymer matrices than larger ones.

It is worthwhile to mention that, in some cases a plot of logarithm of the solubility coefficient vs. the boiling point, the critical temperature or the Lennard-Jones force constant of the gas is useful [36]. In many cases the solubility coefficients correlate well with these variables (Figure 1.6). In addition, the laws of the corresponding states provide a good correlation for gas solubility with the square root of reduced temperatures [7]. A recent investigation by Wang et al. provides solubility coefficients for butane and isobutane in PS and PP (Figure 1.7) [37].

1.3.1 Fickian Diffusion

As described previously, for temperatures well above the T_g of the polymer, diffusion follows Fick's law. Diffusion coefficients are concentration independent. However, if the solubility of the vapor, particularly in the case of organics, is high, thus affecting the molecular interaction between the adjoining chains, the diffusion coefficient exhibits concentration dependence. For a penetrant concentration of 10 wt% in the polymer, the diffusion coefficient can be determined using Equation 1.13:

$$D(C) = D_0 \exp(AC) \qquad (1.13)$$

C is the penetrant concentration and A is the plasticizing parameter related to the Flory-Huggins interaction parameters. For concentrations higher than 10 wt%, replacement of C in the equation by the penetrant activity provides a better correlation [7].

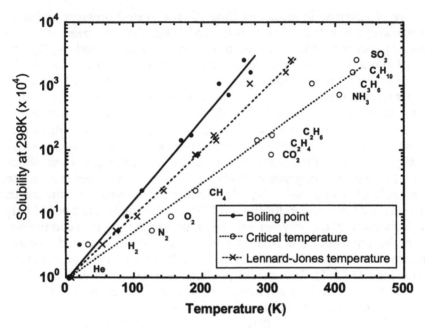

FIGURE 1.6

Solubility of various gases in natural rubber as a function of different temperatures: boiling point, critical temperature, and Lennard-Jones temperature. (Adapted from Van Krevelen, D.W., *Properties of Polymers*, Elsevier, Amsterdam, 1990 [36].)

With respect to polymers at temperatures below their T_g, a plot of concentration vs. the pressure of the gas is frequently convex, due to the fact that sorption in the amorphous region of the polymer obeys Henry's law, while sorption in microvoids obeys the Langmuir isotherm (Equation 1.6). A transport model based on the dual sorption model has been developed and has been applied to diffusion of gasses in glassy polymers [38]. According to this model, at low gas pressures, the apparent diffusivity is less than the actual diffusivity due to absorption of the gas in the dispersed phase. However, at higher pressures the measured solubility and diffusion coefficients are the true values for the matrix.

1.3.2 Non-Fickian Diffusion

This type of behavior is normally observed for penetration of organic vapors in polymers below their T_g, and in some cases also above T_g [7]. The diffusion coefficient becomes time and concentration dependent, and the non-Fickian behavior is observed at higher temperature and with higher penetrant activity. In the case where the swelling effect in the polymer becomes more pronounced at moderate penetrant activities, the diffusion may change from a Fickian to a stress-relaxation-controlled process. At higher activities, stress cracking or crazing may be observed. The term "anomalous diffusion" thus

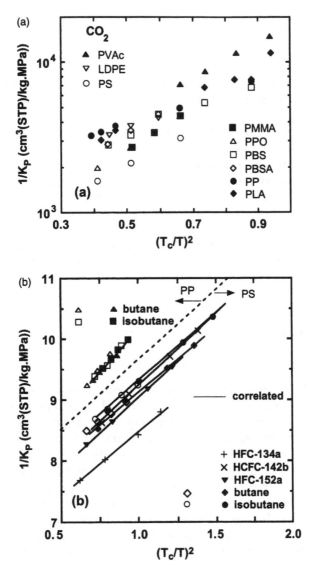

FIGURE 1.7
Henry's law constant as a function of reduced temperature for (a) carbon dioxide in different polymers and (b) several physical foaming agents in PP and PS. (Reproduced from Wang et al. [37] and Sato et al., [68] *Proc. Sympos. Polymer-Supercritical Fluid Systems and Foams*, Tokyo, 2003. With permission.)

refers to cases where a combination of Fickian and non-Fickian diffusion takes place. A quantitative relationship, called the diffusional Deborah number (Equation 1.14), provides a correlation for the types of diffusions that occurs in polymers:

$$\left(Deb\right)_D = \tau_m \left(\frac{x}{D^*}\right) \tag{1.14}$$

where τ_m is the mean relaxation time of the polymer–penetrant system, D^* is the molar average of self diffusion coefficients of polymer and penetrant, and x is a characteristic polymer parameter such as thickness [7]. Therefore, Fickian diffusion occurs when $(Deb)_D \ll 1$ and anomalous diffusion occurs when $(Deb)_D \gg 1$.

In summary, if the penetrant has a low activity where swelling in the polymer becomes small, diffusion follows Fick's law and is concentration independent. However, as the penetrant activities increase, polymer relaxation occurs, resulting in concentration-dependent Fickian diffusion [7]. At much higher activities of penetrant in the polymer, the rates of diffusion and polymer relaxation are very close. As such diffusion coefficients become time and concentration dependent, the gas transport mechanism is termed anomalous diffusion [7].

1.4   Measurement Methods

Experimental determination of diffusion coefficients is carried out by permeability measurements or sorption studies. This section provides a summary of the available techniques and equations used to determine these transport parameters from experimental data. Generally, the permeability measurements and the subsequent determination of diffusion coefficients are used by research groups focusing on membrane transport properties. In the case of the foaming of polymeric materials, the sorption kinetic data is used for determination of the transport parameters in the polymer and their relation to the cellular morphologies created. Hence, this section will provide a brief summary of the membrane techniques used, though the main focus will be on experimental investigation through sorption studies.

1.4.1   Permeation

Permeation measurements are carried out in closed chambers or in a falling stream. Generally, permeation in closed chambers is carried out for transport of condensed vapors, low permeation rates, or studies where continuous analysis is not possible. The second method, based on the falling stream, is normally used for high permeation rates, gas mixtures, or fragile membranes [7]. In the present case, only the closed chamber system will be discussed. Literature provides many good examples and methods for permeation in falling streams [4,7].

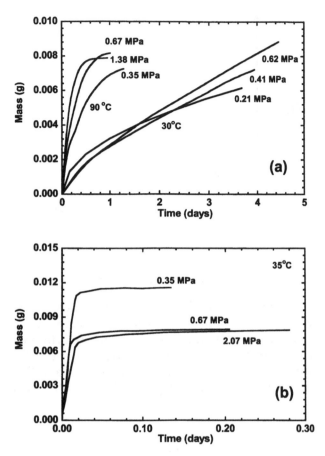

FIGURE 1.8
Sorption kinetics of HFC-134a (a) and carbon dioxide (b) in PS. (From Gendron, R. et al., *Cell. Polym.*, 21, 315, 2002 [39]. With permission.)

1.4.1.1   Mathematical Models

Figure 1.8 provides a representation of the sorption behavior of two physical blowing agents, HFC-134a and CO_2, in PS, thus indicating that different gases have different rates of permeation in the same polymer [39]. However, the ratio of their permeability in the matrix remains constant. Additionally, the ratio of permeability activation energy (E_p), as described by Equation 1.15, and diffusion of a gas remains constant in two different polymers. Therefore, diffusion and permeation coefficients of a gas in a polymer may be estimated when these variables are known for other polymers using the following [40]:

$$P = P_0 \exp\left(\frac{-E_P}{RT}\right) \qquad (1.15)$$

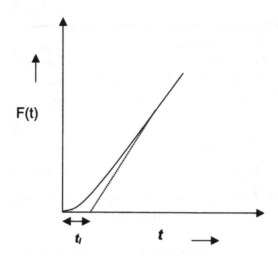

FIGURE 1.9
Typical results of permeation experiment showing the time lag t_l.

P_0 is the pre-exponential factor. If the sorption follows Henry's law (Equation 1.1) and diffusion follows Fick's law (Equation 1.7), where the diffusion coefficient is independent of the gas concentration, the permeability ($P = SD$) is obtained from a closed volume experimental setup. The steady-state permeation rate is thus obtained from Equation 1.5, and the diffusion coefficient is determined from the time lag t_l, as described by Equation 1.16:

$$D = \frac{l^2}{6t_l} \tag{1.16}$$

where l is the film thickness.

Figure 1.9 provides typical permeation resulting from the time lag t_l, assuming that the concentration on the gas downstream is negligible compared to upstream pressure or concentration. The solubility coefficient S is equal to P/D. For various shapes of membranes, details regarding the determination of the diffusion coefficient are provided in the literature [4,5,7].

If the gas sorption volumes are small, errors might be associated with determining transport parameters and therefore correction factors need to be applied [7]:

$$\eta = 0.278 \frac{STAl}{V_{cell}} \tag{1.17}$$

$$\alpha = 0.278 \frac{TAl}{V_{cell}} \tag{1.18}$$

$$S = \frac{PD^{-1}}{1 + 0.38\alpha PD^{-1}}$$ (1.19)

$$P = P(1 + 0.69\alpha S)$$ (1.20)

$$D = D(1 + 0.29\alpha S)$$ (1.21)

In the above equations, η is a defined parameter; S is the true Henry's solubility coefficient; T is the temperature in Kelvin; A and l are the membrane area and thickness, respectively; and V_{cell} is the volume of the downstream chamber. Parameters P, D, and S are the estimated values of the transport parameters through experimental studies. Therefore, plots of P/P, D/D, and S/S vs. η yield the true values of P, D, and S through an iterative procedure [7]. The described approach is valid for η values less than 0.35.

In the case of concentration-dependent diffusion coefficients in closed volume permeation systems, a mean value is determined using Equation 1.22:

$$\frac{l^2}{6} \leq D \leq \frac{l^2}{2}$$ (1.22)

In the absence of data on the dependence of D on concentration, this equation can provide estimation for D from time lag data. A variety of methods have been reviewed in the literature [4,7].

In the case of the transport of gases in glassy polymers, the determination of the permeation coefficient becomes rather difficult, as there are two populations within the polymer matrix that contribute to the diffusion [41]. A model developed by Paul and Koros adequately describes this behavior, as presented by Equation 1.23:

$$\bar{P} = SD\left(1 + \frac{F'K}{1 + bp_1}\right)$$ (1.23)

$$K = \frac{C'_H b}{S}$$ (1.24)

$\bar{P}$ is the permeation coefficient of the gas in the Langmuirian and Henrian sites; F' is the ratio of the diffusion coefficient in the Langmuirian sites to the diffusion coefficient for sorption following Henry's law, and K is determined from the sorption data. The values of F' and D can be measured from steady state permeation data over a range of pressures.

1.4.1.2 Experimental Apparatus

A recent report in the literature provides a review of the experimental techniques available for permeability measurements [42]. However, most constant volume permeation experiments are designed based on the apparatus presented by Koros et al. [43,44]. Using the constant volume method, for example, the gas transport properties of substituted polyetheretherketone (PEEK) were investigated [45]. Recently, a new device for measurement of gas permeation in polymers at high pressure and temperature was developed for oil application, and the permeability, diffusion, and solubility of CO_2, CH_4, Ar, N_2, and He in PE, polyamide 11 and poly(vinylidene fluoride) were measured [46].

Several models have been proposed for cases where the diffusion coefficient is both time and concentration dependent (*anomalous transport*). Most proposed methods require the measurement of time lag both upstream and downstream of the membrane during sorption and desorption experiments [7]. In a recent investigation by Mogri and Paul [47], gas sorption and transport in poly(alkyl (meth)acrylate)s were determined as a function of temperature. In this study, anomalous permeation was also reported for poly(behenyl acrylate) polymer. The transport properties of CH_4 and CO_2 in this study were measured as described previously [43].

Villaluenga and Seoane report on an integral permeation method for the estimation of gas transport properties of linear low-density polyethylene (LLDPE) membrane [48]. The permeability of O_2, CO_2, He, and N_2 in the range from 298 to 348 K in this polymer was measured. The integral permeation method has been reported to provide results for the transport coefficients that are in agreement with the ones obtained by the time lag method.

1.4.2 Sorption

Sorption measurements are generally recommended for exponentially rapid processes. Direct sorption measurements are sensitive enough to describe quantitatively the changes that occur in a polymer as a result of gradual swelling or relaxation, in addition to being able to determine the equilibrium solubility of the gas in the polymer. With respect to non-Fickian transport, permeation experiments are better suited than sorption to the task of following the changes [7]. Further, there are several experimental techniques used in the measurement of gas uptake or desorption by a polymer. These measurements are made using gravimetric methods [49–52]; the pressure–volume–temperature (PVT) method, which measures the change in pressure or volume of the gas in the environment around the polymer [43]; or Raman chemical imaging [53,54]. All of the above-described methods will provide the equilibrium solubility of the gas in the polymer. Diffusion coefficients are then extracted from the sorption data using the available models as described below. Details on the experimental apparatus and setup will follow the mathematical treatment of the data.

1.4.2.1 Mathematical Models

As described earlier, if diffusion follows Fick's law and solubility follows Henry's law, then both D and S will be concentration independent. The equilibrium solubility of the gas (C) in the polymer (Equation 1.1) is represented by normalizing the amount of sorbed gas with respect to the mass of polymer used. Determination of C at a constant temperature and various pressures will yield a linear plot with slope S (solubility coefficient or Henry's law constant). The dual sorption model (Equation 1.3) is used for cases where the plot of C vs. the gas pressure at a constant temperature is concave with respect to the pressure axis as observed for solubility of gases in glassy polymers.

For the determination of gas diffusion coefficients in polymers, there are several models available, which use information obtained during sorption experiments. Among the available models are the half time, initial slope, and moment methods [4,7], and, most recently, the hybrid model [55].

The *half time* method is the simplest technique for determining the diffusion coefficient of a gas in a polymer film of known thickness [56]. However, in applying models involving Fickian diffusion it is important to bear in mind the experimental conditions employed, while ensuring that the following assumptions are met:

1. The polymer film must be of uniform thickness
2. The transport process must be Fickian
3. The sorbate concentration is constant throughout the experiment

The first step in the half time method requires the calculation of the absorption rate, r_m, using Equation 1.25:

$$r_m = \frac{m_p^t - m_p^i}{m_p^\infty - m_p^i} \tag{1.25}$$

where m_p^t is the total weight of the film sample plus sorbate uptake at time t, which is directly measured from the balance; m_p^i is the initial weight of the film; and m_p^∞ is the final (equilibrium) weight of film sample plus sorbate.

The relationship between the absorption rate r_m and D is presented in Equation 1.26. Based upon this equation it can be seen that for each value of r_m there is a corresponding value of Dt/l^2, where D is the diffusion coefficient, t is the time, and l is the film thickness.

$$\frac{-1}{\pi^2} \ln\left(\frac{\left(m_p^t - m_p^i\right)\pi^2}{8\left(m_p^\infty - m_p^i\right)} \right) = \frac{D}{l^2} t \tag{1.26}$$

Hence, by setting r_m equal to 0.25, 0.5, and 0.75 in the above equation, the following relationships (Equations 1.27–1.29) representing D as a function of time can be established. The average for $D_{1/4}$, $D_{1/2}$, and $D_{3/4}$ is taken as the value for the mean diffusion coefficient of the gas in the polymer:

$$0.01228 = \frac{Dt_{1/4}}{l^2} \tag{1.27}$$

$$0.0492 = \frac{Dt_{1/2}}{l^2} \tag{1.28}$$

$$0.1192 = \frac{Dt_{3/4}}{l^2} \tag{1.29}$$

A more detailed description of the use of this model for various geometries in polymer gas systems was provided by Fava [7].

The *initial slope* model also assumes that diffusion follows Fick's law, while sorption follows Henry's law, and as such can be best described by the following equation:

$$\frac{m_p^t}{m_p^\infty} = \frac{4}{l}\left(\frac{D}{\pi}\right)^{\frac{1}{2}} t^{\frac{1}{2}} \tag{1.30}$$

According to this equation, a plot of m_p^t/m_p^∞ vs. $t^{1/2}$ should be linear for all geometries during the early stages of the experiment. Based upon the slope of the initial part of the plot and the film thickness, the diffusion coefficient of the vapor in the polymer can be calculated. It should be noted, however, that the accuracy of this method is poor in comparison to that of the half time method. Initial slopes have been difficult to establish repeatedly for a fast diffusion process, especially the first few points, which affect the slopes significantly. However, when the equilibrium solubility of the vapor in the polymer or the equilibrium amount desorbed is unknown, the value of D determined with the initial slope method can be more accurate than in the case of the half time method.

Felder et al. derived a model in 1975 for the estimation of diffusion coefficients from permeation data [57]. The *moment* method is a variation of this model applied to sorption measurements. Equation 1.31 is solved by numerical integration, followed by application of the results to Equation 1.32:

$$\tau_s = \int_0^\infty \left[1 - \frac{M_t}{M_\infty}\right] dt \tag{1.31}$$

$$D = \frac{h^2}{12\tau_s} \qquad (1.32)$$

This method uses the entire sorption data available, unlike the half time and initial slope methods. Variations of this method for other geometries are well described, once again, by Crank [58].

The hybrid model is a convenient method for the determination of diffusion coefficients from sorption data [55]. This model assumes Fickian transport through the polymer film, and thus the kinetic data is treated based on the following equations:

$$\frac{M_t}{M_\infty} = \varphi(x)f(x) + [1 - \varphi(x)]g(x) \qquad (1.33)$$

$$f(x) = 4\left(\frac{x}{\pi}\right)^{\frac{1}{2}} \qquad (1.34)$$

$$g(x) = 1 - \left(\frac{8}{\pi^2}\right)\exp\left(-\pi^2 x\right) \qquad (1.35)$$

$$\varphi(x) = \frac{1}{1 + \exp\left(\frac{x - a}{b}\right)} \qquad (1.36)$$

$$x = \frac{Dt}{h^2} \qquad (1.37)$$

where $b = 0.001$, $a = 0.05326$, $\phi(x) = 1$ when $x \le 0.05326$, and $\phi(x) = 0$ when $x > 0.05326$. M_t and M_∞ are the mass of gas sorbed at time t and ∞, respectively, h is the film thickness, and D is the diffusion coefficient. The obvious advantage in using the hybrid model is that this model only requires the kinetic data. With this method it is therefore possible to determine the uncoupled values of D and M_∞.

1.4.2.2 Experimental Techniques

Gas sorption experiments in polymers are mainly performed using gravimetric or barometric methods. It has been reported that gravimetric methods for measuring the change in mass of the polymer during sorption and desorption runs are used for subatmospheric pressures, whereas barometric methods are applied to sorption at higher pressures [7]. However, in recent years, with the advancement of gravimetric measurement methods and the interest in the plastics community in investigating the interaction of benign gases with polymers, newer techniques have been reported.

Periodic removal and measurement of polymer mass change as a result of gas uptake, until a constant weight is observed, has been and is still in some cases applied in the measurement of gas sorption in polymers [59]. However, a major source of error during measurement is the escape of the gas while the sample is being removed from the saturation chamber. The magnitude of the error and the corrections needed are very difficult to estimate, as they depend on many factors. In cases where the process is slow because of gas diffusion in the polymer matrix, this method becomes even more unattractive and time consuming. Another experimental technique applied for the determination of solubility is the pressure decay method [43]. In this method, also called the PVT method, the polymer is placed in a reservoir of known volume and changes in the gas pressure, due to uptake by the polymer with respect to a reference cell, are monitored and recorded. A drawback with this method is the need for precise calibrations. Further, the method can only be applied to systems wherein the parameters of the equation of state for the penetrating gas are known.

Gravimetric methods employing an electronic microbalance have been successfully used, though in some cases over a limited temperature and pressure range [60]. A more recent study provides a design in which an *in situ* high-pressure gravimetric technique is used. This technique employs a microbalance (CAHN D110) in order to measure the solubility and diffusivity of gases in polymers in a wide temperature and pressure range [49]. The sample and reference holders in this system are made of quartz and suspended from both arms in the balance. Using this setup, several polymers have been investigated for their interaction with gases, including gas desorption rates from poly(methyl methacrylate) (PMMA)/CO_2 blown foams [52]. A second apparatus for gas sorption measurement in polymers is the magnetic suspension balance manufactured by Rubotherm, Inc. [61]. The polymer samples are placed in a quartz cell and subsequently suspended from a magnet placed in a high-pressure containment jacket [62]. The quartz microbalances are also employed for investigations into polymer gas systems, as reported by Zhang et al. [50]. The reported setup sustains pressures up to 44 atm and temperatures in the range −25 to 150°C. The crystal oscillation frequency is 1 Hz per 10 MHz. Sorption of CO_2 in various polymers has been reported using this method [50].

Raman spectroscopic methods have recently been extended for investigations of gas solubility in polymers as well as bubble nucleation [53]. The reported study uses a laser Raman spectroscopy in conjunction with a high-pressure optical cell. The cell volume was 2 cm³, and the cell was equipped with a 15 mm thick glass window. The incident laser beam was irradiated onto the polymer–gas sample through the object lens and the glass window, allowing the spectrum to be collected.

1.4.2.3 Investigated Polymer–Gas Systems

To date there have been numerous polymer–gas systems investigated, and Table 1.2 provides a list of many cases reported in the last decades that are relevant to foam applications. A review by Stiel and coworkers [85], on the solubility of gases and volatile liquids in PE and polyisobutylene (PIB) at elevated temperatures, also includes an exhaustive list of references that dates back to the 1960s. More recently, papers from Wang et al. [37] on the solubility of condensable gases in PS and PP, and Sato et al. [68] on the solubility of CO_2 in polymers, present very interesting correlations for the Henry's law constants as a function of reduced temperature. Their results are reproduced in Figure 1.7.

Some specific works should however be underlined. The solubility of CO_2 at supercritical conditions (35°C) in poly(vinyl chloride) (PVC), PS, unplasticized poly(vinyl chloride) (UPVC), and flexible poly(vinyl chloride) (FPVC) has been reported using a gravimetric method [49]. The solubility of HFC-134a in PS at 30, 90, and 120°C was determined, and subsequently all diffusion coefficients were determined using the hybrid model.

Solubility and diffusivity data for $PMMA/CO_2$ systems has also been reported using a gravimetric method. The reported data are for CO_2 at sub- and supercritical conditions (temperature range of 0 to 167°C) [62]. This report provides valuable information on the molecular interaction of CO_2 with this polymer and the physical changes as a result of high gas solubility.

The solubility of blowing agents HFC-134a, HCFC-142b, and HFC-152a in PS was measured in the range 75 to 200°C [81]. At a given pressure and temperature it has been reported that the solubility of HFC-134a, HFC-152a, and HCFC-142b increased in PS. The temperature-dependent interaction parameters have been reported in addition to the Henry's law constant for each system.

The interaction of supercritical CO_2 and CO_2 plus cosolvents with PMMA and PMMA/poly(ε-caprolactone) (PCL) blend has been investigated for impregnation of polymers with host molecules such as cholesterol and albumin [86]. The addition of the liquid cosolvent to CO_2 has been reported to enhance the dissolution of solutes and polymers.

A magnetic suspension microbalance was used, and the solubility and diffusivity of ethylene in semicrystalline PE were measured at temperatures up to 80°C and pressures as high as 66 atm [32]. The ethylene diffusion coefficients in polyethylene were found to increase with an increase in temperature. The solubility of ethylene was found to increase with decreasing temperature and to be in good agreement with the solubility determined theoretically using the Sanchez-Lacombe lattice fluid model. This model is very popular for the prediction of gas solubility in polymer; it relies on three independent parameters for each component of the mixture that can be

TABLE 1.2

Reported Polymer–Gas Systems

Gas	Polymer	T/°C	P_{max}/MPa	Ref.
Carbon dioxide	ABS	0–65	6.0	63
(CO_2)	PC	35	10	64
		100–180	10	65
		40	10	66
	PMMA	33–60	10	64
		0–167	6.0	62
		40	10.5	67
		40	10	66
		25–35	2.0	50
	PLA	—	—	68
	FPVC, UPVC	35	6.0	49
	sPS	35	5.5	23
	PBS, PBSA	50–180	20	69
	PCTFE	40	10.5	67
	PDMS (crosslinked)	40	10.5	67
	PTFE, TPX, PI, CPE	40	10	66
	LDPE	150–200	15	25
				50
		110–200	10–18	70
		120–175	12.5	71
	LDPE/TiO_2 composite	150–200	12	25
	PP	150–200	15	72
		160–200	17	73
	HDPE	150–200	15	72
		160–200	18	73
		25–50	4.0	74
	EEA copolymer	150–200	15	72
	PS	65–130	25–45	8
		35–65	7.0	64
		35	50	49
		40	10	66
		150–200	15	72
		100–180	20	75
		100–200	2.0	76
	PVAc	40–100	17.5	76
	PPO, PPO/PS blends	100–200	20	77
	PSF	25–35	2.0	50
	PET	25–85	2.0	50
Nitrogen	LDPE	120–175	12.5	78
(N_2)		150–200	17.0	79
		120–175	12.5	71
		110	10	70
	LDPE (irradiated)	130–192	100	80
	HDPE	120–175	12.5	71
		160–200	15	73
	PP	180–200	18	73
	PS	40–80	70	8
		40–80	17	73
		100–180	17	75

(continued)

TABLE 1.2 (CONTINUED)

Reported Polymer–Gas Systems

Gas	Polymer	T/°C	P_{max}/MPa	Ref.
Argon (Ar)	LDPE	120–175	12.5	71
	HDPE	120–175	12.5	71
HCFC-142b	PS	75–200	3.0	37, 81
HFC-134a	PS	75–200	3.0	37, 81
		30–120	3.3	49
HFC-152a	PS	75–200	3.0	37, 81
HCFC-22	PS, PP	—	—	37, 82
	LDPE, HDPE, PVC, PC, PPO	—	—	82
CFC-11 and CFC-12	PS, PP	—	—	37, 82
CFC-114	PS, LDPE, HDPE, PP, PVC, PC, PPO	—	—	82
Butane	PS	75–200	3.0	37
	PP	165–210	3.0	37
Isobutane	PS	75–200	3.0	37
	PP	165–210	3.0	37
	LDPE	145–200	10–18	70
Methane	LDPE, PS, PIB	100–225	40–70	83
	HDPE	100–225	40–70	83
		25–50	15	74
Ethylene	PE	50–80	6.6	84

determined from PVT data of the pure components. One additional parameter takes into account the interactions between the gas and the polymer, which can vary with temperature [87].

The fundamental transport parameters for acrylonitrile-butadiene-styrene (ABS)/CO_2 systems have been investigated [63]. Solubilities in the temperature range 0 to 65°C and pressures up to 55 atm are reported; values as high as 25 wt% were observed at 0°C and 34 atm. The derived diffusion coefficients from sorption data indicate a favorable interaction between the two components.

The solubility of CO_2 in a number of polymers, such as poly(butylene succinate) (PBS), poly(butylene succinate-co-adipate) (PBSA), poly(vinyl acetate) (PVAc), polylactide (PLA), PP, and LDPE, has been reported by Sato et al. [68]. Results have been presented in terms of solubility coefficient or the Henry's law constant as a function of temperature for each case. Interactions of polyethylene terephthalate (PET), PMMA, and polysulfone with CO_2 have been investigated both above and below the glass transition temperature of the polymers [50].

Data on solubility and diffusivity of methane and CO_2 has been reported for HDPE [74]. The data reported is in the range of 25 to 50°C, at pressures up to 150 atm for methane and 40 atm for CO_2. The solubility results follow Henry's law except at higher pressures of methane.

1.5   Impact on Foam Processing and Foam Properties

A complete dissolution of a physical blowing agent in a polymer or polymer mixture is essential for producing foams with uniform fine-cell structure, and an incomplete dissolution of the blowing agent leads to greater heterogeneity of the porous structure [88] and the appearance of surface defects [89]. Consequently, in order to produce good quality foams, the amount of blowing agent added to a polymer should not exceed its solubility in the polymer before foaming.

1.5.1   Plasticization

First, the presence of a blowing agent in a polymer has a plasticizing effect on the polymer, because blowing agents are small molecules that can be trapped between polymer molecules and reduce their interactions. The reduction of glass transition temperature (T_g) is the direct impact of this effect, and experiments have determined that T_g depression normally has a linear relationship with gas solubility [90,91], in agreement with the model developed by Chow [92]. It is associated with the decrease of solid polymer modulus and the decrease of liquid polymer viscosity [93,94], which affect the bubble growth directly. The modulus reduction is a necessary condition for foaming microcellular foams from solid state in batch processes, since the bubble growth can only occur when the polymer can be easily deformed under the gas pressure created in the nucleated sites. During the foaming process from solid state, bubble growth can be easily frozen since the plasticization effect is decreased, and thus the modulus increases, with the gas continuously phasing out. Such an auto-freezing mechanism is the fundamental reason why such small spherical cells can be easily obtained from batch processes. In addition, the viscosity reduction plays an important role in controlling the bubble growth kinetics during the foaming process from polymer melt, i.e., when the temperature is above the T_g of the neat polymer and the polymer is still allowed to deform. The reduced viscosity favors fast bubble growth; the effect is well described in the newly developed foam-growth models [95].

For semicrystalline polymers, plasticization also delays the crystallization process as shown in Figure 1.10 [96,97]. This effect might be beneficial for the bubble growth occurring in the molten state, since the crystallization of some polymers, such as PE and PP, happens very fast and can easily freeze the expansion of the cells due to the high modulus imparted to the formed crystalline phase.

1.5.2   Solubility vs. Phase Separation

Foaming starts when the dissolved blowing agent is suddenly phased out of the polymer due to the abrupt reduction of its solubility. This is caused

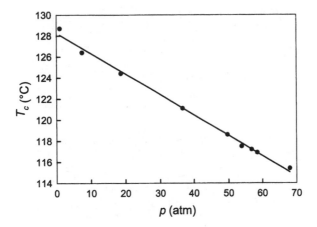

FIGURE 1.10
Variation of crystallization temperature T_c of iPP as a function of CO_2 pressure. The samples were cooled from 200°C at 5°C/min. (From Zhang, Z. et al., *J. Polym. Sci.: Part B: Polym. Phys.*, 41, 1518, 2003 [96]. Reprinted by permission of John Wiley & Sons, Inc.)

by external conditions, such as pressure, temperature, or stress. In other words, the reduction of blowing-agent solubility (called gas solubility) is the driving force for foaming. Theoretically, any changes that can cause a sudden reduction in gas solubility can be used to foam polymers.

The applications of the mechanisms as presented in Equations 1.1 and 1.2 have led to two distinctive foaming techniques. The first one consists in foaming from the polymer in the liquid state (i.e., melt), which is associated with the well-established foam extrusion and injection molding process [98]. The second relies on foaming from a solid polymer, and consists of the batch foaming process used mainly for the production of microcellular foams [99,100]. In the former case, a polymer melt with a dissolved blowing agent is extruded out of the pressurized dies and transferred into an environment with a lower pressure. The associated pressure drop is an important parameter that should be taken into account at the die design stage. It also impacts notably the cell nucleation density and is thus one of the key variables for the continuous production of microcellular foams [101,102]. In the batch foaming process, a polymer with a dissolved blowing agent is heated to a certain temperature under ambient pressure or under a high gas pressure [99,100]. This temperature has been explored for its ability to control the foaming process, as it significantly affects the foam structure [100,103,104].

Another mechanism of reducing gas solubility is to expose a polymer-gas system to a substantial hydrostatic stress, exhibiting a squeezing function. In this case, foaming occurs when a polymer–gas mixture is pressed with a hydrostatic stress of more than 20 MPa under ambient conditions. Figures 1.11 and 1.12 represent the PMMA image foams obtained from this mechanism and the effect of hydrostatic stress magnitude on PMMA foam density [105].

FIGURE 1.11
Photograph of images obtained by pressing PMMA containing 22 wt% CO_2. (From Handa, Y.P. and Zhang, Z., *Cell. Polym.*, 19, 77, 2000 [105]. With permission.)

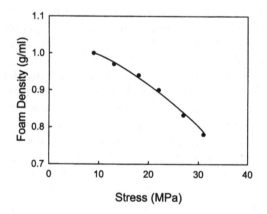

FIGURE 1.12
Density of foams made by pressing PMMA containing 22 wt% CO_2 to various stresses for 5 s using molds at 24°C. (From Handa, Y.P. and Zhang, Z., *Cell. Polym.*, 19, 77, 2000 [105]. With permission.)

1.5.3   Cell Density

While solubility depression is the driving force for foaming, solubility itself plays an important role in affecting the foaming process directly and indirectly. High gas solubility in a polymer means that a large amount of blowing agent can be phased out of the dissolved polymer for favored nucleation and continuous cell growth. Experiments on various systems have showed that cell density, which is dependent on nucleation rate, is important for fine-cell structure and increases with increasing gas content prior to foaming [106–108]. For instance, the cell density of PMMA microcellular foams

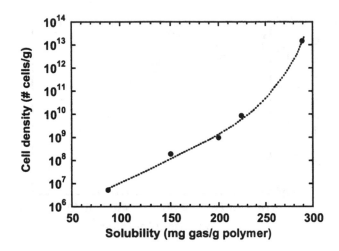

FIGURE 1.13
Cell density as a function of solubility for a PMMA–CO_2 system. (Data from Handa, Y.P. and Zhang, Z., *J. Polym. Sci.: Part B: Polym. Phys.*, 38, 716, 2000 [90].)

obtained from batch process increases monotonically with the initial CO_2 solubility in PMMA (see Figure 1.13), and extremely high cell densities, in the 10^{13} cells/g range, can be achieved when the gas content is superior to 25 wt% [90].

1.5.4 Nucleation and Cell Growth

The presence of sufficient blowing agent is essential to support efficient cell growth. This depends on the gas pressure generated inside the cells, which in turn is dictated by the solubility and diffusivity of the blowing agent in the polymer, as well as its physical properties. Thus, a certain level of solubility is necessary to achieve low foam density.

Diffusivity also affects both nucleation and bubble growth. A high diffusivity favors fast nucleation and thus fine-cell structure, due to a large amount of blowing agent being phased out within a short period. However, this does not help bubble growth, since the blowing agent can easily diffuse out of the cells, losing its blowing power, and escape from the foaming skin. In addition, a high diffusive blowing agent is usually associated with a high gas activity and might easily cause cell rupture. Consequently, smaller molecular blowing agents, such as CO_2, are more suited to generating foams with high cell density and high foam density, while larger molecules, such as fluorocarbons and hydrocarbons, are well adapted for the production of low-density foams with associated low cell density.

Despite several attempts in microcellular foaming to combine small cells with low densities, both goals could not be achieved simultaneously. Obviously the success of such a quest would require that a huge quantity of gas (typically CO_2) be dissolved in the polymer matrix [109]. The high solubility

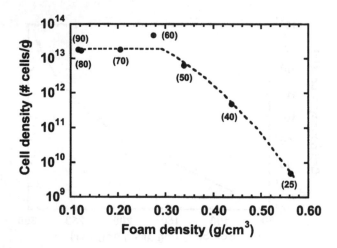

FIGURE 1.14

Foam characteristics (cell density vs. foam density) obtained by foaming a PMMA sample charged with CO_2 and exposed to various temperatures (°C), indicated in parentheses. (Data from Handa, Y.P. and Zhang, Z., *J. Polym. Sci.: Part B: Polym. Phys.*, 38, 716, 2000 [90].)

of carbon dioxide in PMMA reported previously contributes to the surprising results displayed in Figure 1.14, where a foam density close to 0.1 g/cm^3 matches a cell density greater than 10^{13} cells/g.

Although low diffusivity is preferred for cell growth, a fast diffusion process is necessary for dissolving a blowing agent into the polymer rapidly. As in foaming processes like extrusion and injection molding, where the residence time can be relatively short, this difficulty is overcome by increasing the temperature at the mixing zones, since diffusivity increases exponentially with temperature. High mixing temperatures are associated with low gas solubility in the polymer melt, which exponentially decreases with temperature. This negative impact can be circumvented, fortunately, with adequate cooling in the subsequent zones, which has the reverse impact on solubility.

1.5.5   Diffusion vs. Applications

When the foaming process is complete, blowing agents may leave or stay in the polymers, depending on their diffusivity at the environment temperature. While some blowing agents, such as CO_2 and nitrogen, can escape the foams quickly, others, such as hydrocarbons and fluorocarbons, may leave the foams slowly. Depending on the processes and applications enacted after foaming, the diffusion of blowing agents from foams needs to be properly addressed.

For many packaging foams, the moderately slow permeation of hydrocarbons has a severe drawback, and a quarantine period should be provided for foam aging, in order to get rid of flammable hydrocarbons before foams are shipped to consumers; or foams should be post-processed, like the PS sheets that are thermoformed into food trays. This diffusion necessitates a huge storage capacity, with associated fire hazard [110].

Permeability also impacts the dimensional stability of foams made from low-modulus resins, such as polyolefin foams. The hydrocarbons used in the foaming process permeate out of the foam 5 to 60 times faster than the rate of air diffusing into the cell, which makes the pressure inside the cells lower than the atmosphere. This causes a temporary collapse or shrinkage during short-term aging [111]. Permeation modifiers, like fatty acid esters or amides, should therefore be incorporated to regulate the foaming agent–air interchange in the extruded foam. It has been proposed that the mechanism of these permeation inhibitors relies on their blooming at the interface of the cell and polymer matrix and thus forming a monolayer that impedes the permeation of larger molecules, as compared to air [110].

For the foam boards used in thermal insulation applications, such as those obtained from polystyrene extrusion foaming, conductivity of the gas trapped inside the cell remains highly significant, compared to the other mechanisms that control the heat transfer properties, such as solid (polymer) conductivity and radiative (infrared) transmission [112]. Although a maximum insulation property can be obtained by elimination of the conductive gas through a new technology called *vacuum insulated panel* (VIP), the appropriate choice for a low-conductivity gas remains critical for conventional commercial foam boards. In addition to its low thermal conductivity properties, the gas should remain within the panel as long as possible (low permeability), before being replaced gradually by air, the latter exhibiting poor resistance to heat transfer (k_{air} = 26.24 mW/mK at 25°C). Originally, panels were foamed with CFC-12 (k = 9.6 mW/mK at 25°C); this was gradually replaced by HCFC-142b (11.5 mW/mK) after the ban of the ozone-depleting substances. Searches for a viable alternative have identified HFC-134a (13.6 mW/mK) and CO_2 (16.6 mW/mK) as potential replacements [39]. Unfortunately, despite being very attractive, carbon dioxide suffers from higher permeability in PS, which would induce rapid aging of the insulation panels with low long-term performance.

1.6   Retrograde Behavior

Most polymer–gas systems have a glass transition temperature at a given gas pressure or gas concentration, which decreases linearly with gas pressure or gas concentration in polymers. Some special polymer–gas systems, however, have two T_g values at a given gas pressure. One of those T_g values is the normally expected plasticized T_g corresponding to the glass-to-rubber transition; the other T_g is an unusual one, corresponding to the glass-to-rubber transition known as retrograde vitrification, as shown in Figure 1.15

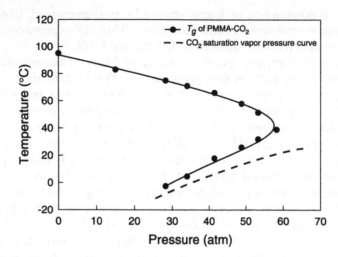

FIGURE 1.15
Glass transition temperature of PMMA/CO_2 as a function of the gas pressure. The dashed curve represents the vapor–liquid equilibrium boundary for CO_2. (From Handa, Y.P. and Zhang, Z., *J. Polym. Sci.: Part B: Polym. Phys.*, 38, 716, 2000 [90]. Reprinted by permission of John Wiley & Sons, Inc.)

[90]. The general phenomenon of retrograde vitrification was observed during theoretical calculations of the solubility of CO_2 in various polymers over wide ranges of temperature and pressure [6,93]. It was found that for polymers with T_g values not too far away from the critical temperature of CO_2 ($T_c = 31.1°C$), the solubility of CO_2 at low temperatures is quite large, and that the solubility changes quite rapidly with temperature. Thus, at low temperatures, so much gas dissolves in the polymer that the system goes into a rubbery state. On heating at a constant pressure, the polymer loses the gas quite rapidly, such that it undergoes the retrograde transition of a rubber turning into a glass on heating. The resulting glass still has a considerable amount of gas dissolved in it, so that on continued heating below the glass transition temperature of the neat polymer, the polymer enters the rubbery state again. The system thus undergoes two glass transitions at a given gas pressure.

Retrograde vitrification was observed experimentally in two polymer–gas systems, namely, PMMA/CO_2 and poly(ethyl methacrylate)/CO_2, using creep compliance measurement and heat capacity measurement using high-pressure DSC [62,90,91,93]. The phenomenon was also observed from diffusivity measurement using a high-pressure balance [62]. The findings are associated with unusually high gas solubility, which is the fundamental reason for retrograde vitrification. As is shown in Figure 1.16, the CO_2 solubility in PMMA is unusually higher than the van't Hoff plot when the temperature is low. Note that the gas solubility of polymer–gas systems normally follows the van't Hoff plot, which can be used to calculate the heat of sorption [62]. Such a high solubility is due to the strong molecular

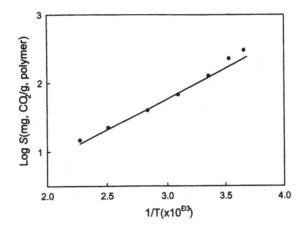

FIGURE 1.16
Solubility of CO_2 in PMMA as a function of temperature. Solid line represents van't Hoff plot. (From Handa, Y.P. and Zhang, Z., *J. Polym. Sci.: Part B: Polym. Phys.*, 38, 716, 2000 [90]. Reprinted by permission of John Wiley & Sons, Inc.)

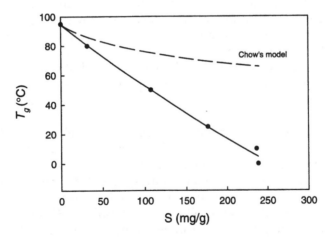

FIGURE 1.17
Glass transition temperature of PMMA/CO_2 as a function of solubility, in comparison with predictions from Chow's model. (From Handa, Y.P. et al., *Cell. Polym.*, 20, 1, 2001 [62]. With permission.)

interactions between CO_2 and PMMA, which keeps the polymer in a rubbery state at low temperature. This strong interaction also causes an unusually strong plasticizing effect of CO_2 on the polymer, as shown in Figure 1.17. A reduction in the T_g of the PMMA/CO_2 system is observed, as a function of gas solubility [62]. The T_g depression of PMMA is much greater than that predicated by Chow's equation which is followed by many polymer–gas systems [92].

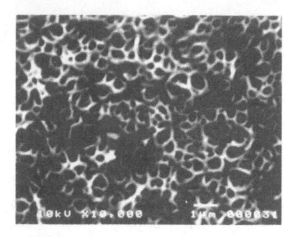

FIGURE 1.18
SEM microphotograph of PMMA foam made by saturating the polymer at 0°C and 34 atm and foaming at 80°C.

In addition to an unusually high solubility, retrograde vitrification is associated with an unusual diffusion behavior. It was found that, under a constant pressure, CO_2 diffusion might be faster at a lower temperature than at a higher temperature. For example, under 34 atm, the diffusion of CO_2 in PMMA at 0°C is faster than that at 10°C [62]. This phenomenon is in conflict with the trend observed in most polymer systems, but is in agreement with the fact that gas diffuses faster in rubbery polymers than in glassy polymers. PMMA is plasticized to a rubbery state at 0°C but stays in a glassy state at 10°C.

Since unusually high gas solubility is associated with retrograde vitrification, foams with unusually high cell density and small cell size can be achieved from the involved polymers. Figure 1.18 shows the ultramicrocellular foam of PMMA obtained by saturating PMMA with CO_2 at 0°C and 34 atm. Heating the system at 80°C under ambient pressure, cell density has reached 4.4×10^{13} cells/g, and cell size is as small as 0.35 μm. Foams with a variety of morphologies, in fact, can be achieved by utilizing the retrograde envelope shown in Figure 1.15 [90].

1.7 Conclusions

As presented in this chapter, bringing fluids into contact with polymers can drastically modify the fundamental physical properties of the latter — i.e., the glass transition temperature for amorphous polymers and the melting point for the semicrystalline ones, these characteristics usually probed with high pressure DSC. The extent of this modification is primarily controlled by the solubility of the gas, which can be investigated using several experimental techniques, such as gravimetric, permeation, pressure decay, and Raman imaging methods. Solubility is affected by several conditions: temperature, pressure, crystallinity fraction, and most importantly, the interaction between the gas and the polymer. Coupled with mathematical models, the transient sorption data can also provide information on the diffusivity of the gas within the polymer, a highly relevant parameter when it comes to process implications or product applications.

Most recently, supercritical fluids have received considerable attention in the study of polymer–gas systems. However, the extent of solubility of the fluid in the polymer is not only dependent on the state of the gas but more importantly on the association of the polymer with the gas itself. As presented in this chapter, conditioning PMMA with CO_2 at subambient conditions can result in morphologies with cell sizes at the submicron level. This example illustrates that when a polymer and a gas exhibit a strong interaction, the lower depressions of the T_g observed in addition to the increase in the relaxation times leads to finer morphologies following the phase separation step. A better understanding of the gas–polymer interaction, as well as a thorough mapping of the solubility and diffusivity properties of the mixture, obviously provides opportunities in developing new foam technologies, or improving existing ones.

Abbreviations

ABS: Acrylonitrile-butadiene-styrene copolymer
CFC: Chlorofluorocarbon
CPE: Chlorinated polyethylene
EEA: Ethylene–ethyl–acrylate copolymer
HDPE: High-density polyethylene
FPVC: Flexible poly(vinyl chloride)
HCFC: Hydrochlorofluorocarbon
HFC: Hydrofluorocarbon
iPP: Isotactic polypropylene
LDPE: Low-density polyethylene
PBS: Poly(butylene succinate)
PBSA: Poly(butylene succinate-co-adipate)
PC: Polycarbonate
PCL: Poly(ε-caprolactone)
PCTFE: Poly(chlorotrifluoroethylene)

PDMS: Poly(dimethylsiloxane)

PE: Polyethylene

PEEK: Polyetheretherketone

PEMA: Poly(ethyl methacrylate)

PET: Polyethylene terephthalate

PI: Polyimide

PIB: Polyisobutylene

PLA: Polylactide

PMMA: Poly(methyl methacrylate)

PP: Polypropylene

PPO: Poly(phenylene oxide)

PS: Polystyrene

PSF: Polysulfone

PTFE: Teflon

PVAc: Poly(vinyl acetate)

PVC: Poly(vinyl chloride)

sPS: Syndiotactic polystyrene

TPX: Poly(4-methyl-1-pentene)

UPVC: Unplasticized poly(vinyl chloride)

VIP: Vacuum insulation panel

References

1. Krause, B., Diekmann, K., van der Vegt, N.F.A., and Wessling, M., Open nanoporous morphologies from polymeric blends by carbon dioxide foaming, *Macromolecules*, 35, 1738, 2002.
2. Koros, W.J., Chan, A.H., and Paul, R.D., Sorption and transport of various gases in polycarbonate, *J. Membr. Sci.*, 2, 165, 1977.
3. Frisch, K.C. and Klempner, D., Introduction, in *Polymer Foams*, Klempner, D. and Frisch, K.C., Eds., Hanser Publishers, Munich, 1991, chap. 1.
4. Crank, J. and Park, G.S., Methods of measurement, in *Diffusion in Polymers*, Crank, J. and Park, G.S., Eds., Academic Press, London, 1968, chap. 1.
5. Petropoulos, J.H., Mechanisms and theories for sorption and diffusion of gases in polymers, in *Polymer Gas Separation Membranes*, Paul, R.D. and Yampol'skii, Y.P., Eds., CRC Press, Boca Raton, 1994, chap. 2.
6. Kalospiros, N.S. and Paulaitis, M.E., Molecular thermodynamic model for solvent-induced glass transitions in polymer-supercritical fluid systems, *Chem. Eng. Sci.*, 49, 659, 1994.
7. Felder, R.M. and Huvard, G.S., Permeation, diffusion, and sorption of gases and vapors, in *Methods of Experimental Physics, Vol. 16: Polymers, Part C: Physical Properties*, Fava, R.A., Ed., Academic Press, New York, 1980, chap. 17.

8. Hilic, S. et al., Simultaneous measurement of the solubility of nitrogen and carbon dioxide in polystyrene and of the associated polymer swelling, *J. Polym. Sci. Part B: Polym. Phys.*, 39, 2063, 2001.
9. Pauly, S., Permeability and diffusion data, in *Polymer Handbook*, 4th ed., Brandrup, J., Immergut, E.H., and Grulke, E.A., Eds., John Wiley & Sons, New York, 1999, chap. 6.
10. Weinkauf, D.H. and Paul, R.D., Effects of structural order on barrier properties, in *Barrier Polymers and Structures*, Koros, W.J., Ed., American Chemical Society, Washington, D.C., 1990, chap. 3.
11. Michaels, A.S. and Bixler, H.J., Solubility of gases in polyethylene, *J. Polym. Sci.*, 50, 393, 1961.
12. Griskey, R.G., Behavior of gases in structural foam during modelling, *Modern Plast.*, 54, 72, 1977.
13. Fleming, G.K. and Koras, W., Dilation of polymers by sorption of carbon dioxide at elevated pressures. 1. Silicone rubber and unconditioned polycarbonate, *Macromolecules*, 19, 2285, 1986.
14. Zhou, Z., Schultze, J.D., and Springer, J., Gas sorption in poly(butylene terephthalate). II. Influence of crystallinity and molecular orientation, *J. Appl. Polym. Sci.*, 47, 13, 1993.
15. Chan, A.H. and Paul, D.R., Influence of history on the gas sorption, theoretical and mechanical properties of glassy polycarbonate, *J. Appl. Polym. Sci.*, 24, 1539, 1979.
16. Puleo, A.C., Paul, D.R., and Wong, P.K., Gas sorption and transport in semicrystalline poly(4-methyl-1-pentane), *Polymer*, 30, 1357, 1989.
17. Ciora, R.J. and Magill, J.H., Novel rolltruded films. 4. Gas separation characteristics of rolltruded isotactic polypropylene, *Polymer*, 35, 949, 1994.
18. Brolly, J.B., Bower, D.I., and Ward, I.M., Diffusion and sorption of CO_2 in poly(ethylene terephthalate) and poly(ethylene naphthalate), *J. Polym. Sci. Part B: Polym. Phys.*, 34, 769, 1996.
19. Ciora, R.J. and Magill, J.H., Separation of small molecules using novel rolltruded membranes. I. Apparatus and preliminary results, *J. Polym. Sci. Part B: Polym. Phys.*, 30, 1035, 1992.
20. Horas, J.A. and Rizzotto, M.G.H., Gas diffusion in partially crystalline polymers. Part I: Concentration dependence, *J. Polym. Sci. Part B: Polym. Phys.*, 34, 1541, 1996.
21. Hirose, T., Mizoguchi, K., and Kamiya, Y., Sorption and transport of CO_2 in poly(ethylene terphthalate) crystallized by sorption of high pressure CO_2, *J. Appl. Polym. Sci.*, 37, 1513, 1989.
22. Ciora, R.J. and Magill, J.H., Novel rolltruded films. II. Effect of draw ratio on the gas transport properties of rolltruded isotactic poly(propylene), *J. Polym. Sci. Part B: Polym. Phys.*, 32, 305, 1994.
23. Handa, Y.P. et al., Gas solubility in and foamability of various forms of syndiotactic polystyrene, *Cell. Polym.*, 20, 4, 2001.
24. Ciora, R.J. and Magill, J.H., Novel rolltruded membranes. III. The effect of processing temperature on the gas transport properties of isotactic polypropylene, *J. Appl. Polym. Sci.*, 58, 1021, 1995.
25. Areerat, S. et al., Solubility of carbon dioxide in polyethylene/titanium dioxide composite under high pressure and temperature, *J. Appl. Polym. Sci.*, 86, 282, 2002.

26. Alexander, M. and Dubois, P., Polymer-layered silicate nanocomposites: preparation, properties and use of a new class of materials, *Mater. Sci. Eng.*, 28, 1, 2000.
27. Nam, P.H. et al., Foam processing and cellular structure of polypropylene/clay nanocomposites, *Polym. Eng. Sci.*, 42, 1907, 2002.
28. Fujimoto, Y. et al., Well-controlled biodegradable nanocomposite foams: from microcellular to nanocellular, *Macromol. Rapid Commun.*, 24, 457, 2003.
29. Mitsunaga, M. et al., Intercalated polycarbonate/clay nanocomposites: nanostructure control and foam processing, *Macromol. Rapid Commun.*, 24, 457, 2003.
30. Treybal, R.E., *Mass-Transfer Operations*, 3rd ed., McGraw-Hill, New York, 1980, chap. 1.
31. Limm, W. and Hollfield, H.C., Modeling of additive diffusion in polyolefins, *Food Add. Contam.*, 13, 949, 1996.
32. Kiparissides, C. et al., Experimental and theoretical investigation of solubility and diffusion of ethylene in semicrystalline PE at elevated pressures and temperatures, *J. Appl. Polym. Sci.*, 87, 953, 2003.
33. Wilks, B. and Rezac, M.E., Impact of average free-volume element size on transport in sterioisomers of polynorborene. II. Impact of temperature and solubility, *J. Polym. Sci. Part B: Polym. Phys.*, 41, 1939, 2003.
34. Vrentas, J.S. and Vrentas, C.M., Viscoelastic diffusion, *J. Polym. Sci. Part B: Polym. Phys.*, 39, 1529, 2001.
35. Neergaard, J., Hassager, O., and Szabo, P., Molecular model for solubility of gases in flexible polymers, *J. Polym. Sci. Part B: Polym. Phys.*, 41, 701, 2003.
36. Van Krevelen, D.W., *Properties of Polymers*, 3rd ed., Elsevier, Amsterdam, 1990, 540.
37. Wang, M. et al., Solubility of condensable gases in polymers, in *Proc. Symp. Polym.-Supercrit. Fluid Syst. Foams (P-(SF)²)*, Tokyo, Japan, 2003, 141.
38. Paul, D.R., Gas sorption and transport in glassy polymers, *Ber. Bunsenges. Phys. Chem.*, 83, 249, 1979.
39. Gendron, R. et al., Foam extrusion of polystyrene blown with HFC-134a, *Cell. Polym.*, 21, 315, 2002.
40. Mulder, M., Introduction and transport in membranes, in *Basic Principles of Membrane Technology*, Kluwer Academic Publishers, the Netherlands, 1991, chaps. 1 and 5.
41. Paul, R.D. and Koros, W.J., Design considerations for measurement of gas sorption in polymers by pressure decay, *J. Polym. Sci. Part B: Polym. Phys.*, 14, 675, 1976.
42. Flaconnèche, B., Martin, J., and Klopffer, M.H., Transport of gases in polymers, *Oil Gas Sci. Technol. Rev. IFP*, 56, 261, 2001.
43. Koros, W.J., Paul, D.R., and Rocha, A.A., Carbon dioxide sorption and transport in polycarbonate, *J. Polym. Sci. Part B: Polym. Phys.*, 14, 687, 1976.
44. Koros, W.J., Simplified analysis of gas/polymer selective solubility behavior, *J. Polym. Sci. Part B: Polym. Phys.*, 23, 1611, 1985.
45. Handa, Y.P., Roovers, J., and Moulinié, P., Gas transport properties of PEEKs, *J. Polym. Sci. Part B: Polym. Phys.*, 35, 2355, 1997.
46. Flaconnèche, B., Martin, J., and Klopffer, M.H., Permeability, diffusion and solubility of gases in polyethylene, polyamide 11 and poly(vinylidene fluoride) *Oil Gas Sci. Technol. Rev. IFP*, 56, 245, 2001.
47. Mogri, Z. and Paul, R.D., Gas sorption and transport in poly(alkyl (meth)acrylate)s. II. Sorption and diffusion properties, *Polymer*, 42, 7781, 2001.

48. Villaluenga, J.P.G. and Seoane, B., Experimental estimation of gas-transport properties of linear low-density polyethylene membranes by an integral permeation method, *J. Appl. Polym. Sci.*, 82, 3013, 2001.
49. Wong, B., Zhang, Z., and Handa, Y.P., High-precision gravimetric technique for determining the solubility and diffusivity of gases in polymers, *J. Polym. Sci. Part B: Polym. Phys.*, 36, 2025, 1998.
50. Zhang, C. et al., Glassy polymer-sorption phenomena measured with a quartz crystal microbalance, *J. Polym. Sci. Part B: Polym. Phys.*, 41, 2109, 2003.
51. Wolf, C.J., Transport and solubility of fluids into polyphenylene sulfide (PPS), *J. Appl. Polym. Sci.*, 90, 615, 2003.
52. Nawaby, A.V. and Handa, Y.P., The second expansion cycle in ultramicrocellular foams, *Cell. Polym.*, 22, 260, 2003.
53. Takahashi, M. et al., Raman spectroscopic investigation of the phase behavior and phase transition in a poly(methyl methacrylate)-carbon dioxide system, *J. Polym. Sci. Part B: Polym. Phys.*, 41, 2214, 2003.
54. Garcia, D., Nelson, M.P., and Treado, P.J., Applications of Raman chemical imaging to polymeric systems, *Polym. Prepr.*, 43, 1271, 2002.
55. Balik, C.M., On the extraction of diffusion coefficients from gravimetric data from sorption of small molecules by polymer thin films, *Macromolecules*, 29, 3025, 1996.
56. Hernandez-Muñoz, P., Gavara, R., and Hernandez, R.J., Evaluation of solubility and diffusion coefficients in polymer film-vapor systems by sorption experiments, *J. Membr. Sci.*, 154, 195, 1999.
57. Felder, R.M. et al., A method for the dynamic measurement of diffusivities of gases in polymers, *J. Appl. Polym. Sci.*, 19, 3193, 1975.
58. Crank, J., *The Mathematics of Diffusion*, 2nd ed., Oxford University Press, London, 1975.
59. Berens, A.R. and Huvard, G.S., Interaction of polymers with near-critical carbon dioxide, *ACS Symp. Ser.*, 406, 207, 1989.
60. Kamiya, Y. et al., Gravimetric study of high-pressure sorption of gases in polymers, *J. Polym. Sci. Part B: Polym. Phys.*, 24, 1525, 1986.
61. Rubotherm Magnetic Suspension Balances, Universitätsstrasse 142, 44799 Bochum, Germany, http://www.rubotherm.de.
62. Handa, Y.P., Zhang, Z., and Wong, B., Solubility, diffusivity, and retrograde vitrification in PMMA-CO_2 and development of sub-micron cellular structures, *Cell. Polym.*, 20, 1, 2001.
63. Nawaby, A.V. and Handa, Y.P., Solubility, diffusivity, and retrograde vitrification in ABS-CO_2 system and preparation of nanofoams, in *Proc. Symp. Polym.-Supercrit. Fluid Syst. Foams (P-(SF)²)*, Tokyo, Japan, 2003, 36.
64. Wissinger, R.G. and Paulaitis, M.E., Swelling and sorption in polymer-CO_2 mixtures at elevated pressures, *J. Polym. Sci. Part B: Polym. Phys.*, 25, 2497, 1987.
65. Handa, Y.P. and Zhang, Z., Sorption, diffusion, and dilation in linear and branched polycarbonate-CO_2 systems in relation to solid state processing of microcellular foams, *Cell. Polym.*, 21, 221, 2002.
66. Aubert, J.H., Solubility of carbon dioxide in polymers by the quartz crystal microbalance technique, *J. Supercrit. Fluids*, 11, 163, 1998.
67. Webb, K.F. and Teja, A.S., Solubility and diffusion of carbon dioxide in polymers, *Fluid Phase Equilib.*, 158–160, 1029, 1999.

68. Sato, Y., Takishima, S., and Masuoka, H., Solubility of carbon dioxide in polymers, in *Proc. Symp. Polym.-Supercrit. Fluid Syst. Foams (P-(SF)²)*, Tokyo, Japan, 2003, 139.

69. Sato, Y., Solubility and diffusion coefficient of carbon dioxide in biodegradable polymers, *Ind. Eng. Chem. Res.*, 39, 4813, 2000.

70. Chaudhary, B.I. and Johns, A.I., Solubilities of nitrogen, isobutane and carbon dioxide in polyethylene, *J. Cell. Plas.*, 34, 312, 1998.

71. Ruengphrathuengsuka, W., Bubble nucleation and growth dynamics in polymer melts, Ph.D. thesis, Texas A&M University, 1992.

72. Areerat, S., Solubility, diffusion coefficient and viscosity in polymer/CO_2 systems, Ph.D. thesis, Kyoto University, 2002.

73. Sato, Y., Solubilities and diffusion coefficients of carbon dioxide and nitrogen in polypropylene, high-density polyethylene, and polystyrene under high pressures and temperatures, *Fluid Phase Equilib.*, 162, 261, 1999.

74. Von Solms, N. et al., Direct measurement of gas solubilities in polymers with a high-pressure microbalance, *J. Appl. Polym. Sci.*, 91, 1476, 2004.

75. Sato, Y. et al., Solubilities of carbon dioxide and nitrogen in polystyrene under high temperature and pressure, *Fluid Phase Equilib.*, 125, 129, 1996.

76. Sato, Y. et al., Solubilities and diffusion coefficients of carbon dioxide in poly(vinyl acetate) and polystyrene, *J. Supercrit. Fluids*, 19, 187, 2001.

77. Sato, Y. et al., Solubility of carbon dioxide in PPO and PPO/PS blends, *Fluid Phase Equilib.*, 194–197, 847, 2002.

78. Lee, J.G. and Flumerfelt, R.W., Nitrogen solubilities in low-density polyethylene at high temperatures and high pressures, *J. Appl. Polym. Sci.*, 58, 2213, 1995.

79. Dagli, S.S., Staats-Westover, R.F., and Biesenberger, J.A., Measurements of gas solubility and relationship to extruded foam properties: a new apparatus and some data on nitrogen in polyolefins, *Proc. ANTEC '89*, 889, 1989.

80. Atkinson, E.B., The solubility of nitrogen in polyethylene above the crystalline melting point at high pressure, *J. Polym. Sci.: Polym. Phys. Ed.*, 15, 795, 1977.

81. Sato, Y. et al., Solubility of hydrofluorocarbon (HFC-134a, HFC-152a) and hydrochlorofluorocarbon (HCFC-142b) blowing agents in polystyrene, *Polym. Eng. Sci.*, 40, 1369, 2000.

82. Gorski, R.A., Ramsey, R.B., and Dishart, K.T., Physical properties of blowing agent polymer systems. I. Solubility of fluorocarbon blowing agents in thermoplastic resins, *J. Cell. Plast.*, 22, 21, 1986.

83. Lundberg, J.L. and Mooney, E.J., Diffusion and solubility of methane in polyisobutylene, *J. Polym. Sci. Part A-2*, 7, 947, 1969.

84. Kiparissides, C. et al., Experimental and theoretical investigation of solubility and diffusivity of ethylene in semicrystalline PE at elevated pressures and temperatures, *J. Appl. Polym. Sci.*, 87, 953, 2003.

85. Stiel, L.I. et al., The solubility of gases and volatile liquids in polyethylene and polyisobutylene at elevated temperatures, *J. Appl. Polym. Sci.*, 30, 1145, 1985.

86. Domingo, C. et al., Behavior of poly(methyl methacrylate)-based systems in supercritical CO_2 and CO_2 plus cosolvent: solubility measurements and process assessment, *J. Appl. Polym. Sci.*, 90, 3652, 2003.

87. De Angelis, M.G. et al., Hydrocarbon and fluorocarbon solubility and dilation in poly(dimethylsiloxane): comparison of experimental data with predictions of the Sanchez-Lacombe equation of state, *J. Polym. Sci. Part B: Polym. Phys.* 37, 3011, 1999.

88. Gendron, R. et al., Foam extrusion of PS blown with a mixture of HFC-134a and isopropanol, *Cell. Polym.*, 23, 1, 2004.

89. Markov, A.V. et al., Study of the foaming of polymer melts, *Plasticheskie Massy*, 25, T68–T71, 1998.

90. Handa, Y.P. and Zhang, Z., A new technique for measuring retrograde vitrification in polymer-gas systems and for making ultramicrocellular foams from the retrograde phase, *J. Polym. Sci. Part B: Polym. Phys.*, 38, 716, 2000.

91. Zhang, Z. and Handa, Y.P., An *in situ* study of plasticization of polymers by high pressure gases, *J. Polym. Sci. Part B: Polym. Phys.*, 36, 977, 1998.

92. Chow, T.S., Molecular interpretation of the glass transition temperature of polymer-diluent systems, *Macromolecules*, 13, 362, 1980.

93. Condo, P.D. and Johnston, K.P., Retrograde vitrification of polymers, with compressed fluid diluents: experimental confirmation, *Macromolecules*, 25, 6720, 1992.

94. Gendron, R. and Daigneault, L.E., Rheology of thermoplastic foam extrusion process, in *Foam Extrusion: Principles and Practice*, Lee, S.-T., Ed., Technomic Publishing Company, Lancaster, PA, 2000, chap. 3.

95. Ramesh, N.S., Foam growth in polymers, in *Foam Extrusion: Principles and Practice*, Lee, S.-T., Ed., Technomic Publishing Company, Lancaster, PA, 2000, chap. 5.

96. Zhang, Z., Nawaby, A.V., and Day, M., CO_2-delayed crystallization of isotactic polypropylene: A kinetic study, *J. Polym. Sci. Part B: Polym. Phys.*, 41, 1518, 2003.

97. Takada, M., Tanigaki, M., and Ohshima, M., Effects of CO_2 on crystallization kinetics of polypropylene, *Polym. Eng. Sci.*, 41, 1938, 2001.

98. Throne, J.L., *Thermoplastic Foams*, Sherwood Publishers, Hinckley, Ohio, 1996.

99. Ramesh, N.S., Rasmussen, D.H., and Campbell, G.A., Numerical and experimental studies of bubble growth during the microcellular foaming process, *Polym. Eng. Sci.*, 31, 1657, 1991.

100. Kumar, V. and Weller, J., Production of microcellular polycarbonate using carbon dioxide for bubble nucleation, *J. Eng. Ind.*, 116, 413, 1994.

101. Park, C.B., Baldwin, D.F., and Suh, N.P., Effect of the pressure drop rate on cell nucleation in continuous processing of microcellular polymers, *Polym. Eng. Sci.*, 35, 432, 1995.

102. Dey, S.K., Jacob, C., and Xanthos, M., Inert-gas extrusion of rigid PVC foams, *J. Vinyl Addit. Technol.*, 2, 48, 1996.

103. Ramesh, N.S. et al., An experimental study on the nucleation of microcellular foams in high impact polystyrene, *Proc. ANTEC '92*, 1078, 1992.

104. Handa, Y.P. et al., Some thermodynamic and kinetic properties of the system PETG-CO_2 and morphological characteristics of the CO_2-blown foams, *Polym. Eng. Sci.*, 39, 55, 1999.

105. Handa, Y.P. and Zhang, Z., A novel stress-induced nucleation and foaming process and its applications in making homogeneous foams, anisotropic foams, and multilayered foams, *Cell. Polym.*, 19, 77, 2000.

106. Park, C.B. and Cheung, L.K., A study of cell nucleation in the extrusion of polypropylene foams, *Polym. Eng. Sci.*, 37, 1, 1997.

107. Goel, S.K. and Beckman, E.J., Generation of microcellular polymeric foams using supercritical carbon dioxide. I. Effect of pressure and temperature on nucleation, *Polym. Eng. Sci.*, 34, 1137, 1994.

108. Zipfel, L. et al., HFC-365mfc and HFC-245fa progress in application of new HFC blowing agents, *J. Cell. Plast.*, 34, 511, 1998.

42   *Thermoplastic Foam Processing: Principles and Development*

109. Beckman, E.J., Generation of microcellular foam using carbon dioxide, in *Proceedings of Foamplas 2000 – Third Worldwide Conference on Foamed Plastic Markets and Technology,* Schotland Business Research, Inc., Skillman, NJ, 2000, 85.
110. Park, C.P., An overview of polyolefin foams: opportunities, challenges and recent developments, in *Proc. Foams '99 — First Int. Conf. Thermoplastic Foams,* Society of Plastics Engineers, 1999, 61.
111. Yang, C.-T., Lee, K.L., and Lee, S.-T., Dimensional stability of LDPE foams: modeling and experiments, *J. Cell. Plast.,* 38, 113, 2002.
112. Almanza, O.A., Rodriguez-Pérez, M.A., and de Saja, J.A., Prediction of the radiation term in the thermal conductivity of crosslinked closed cell polyolefin foams, *J. Poly. Sci. Part B: Polym. Phys.,* 38, 993, 2000.

2

Rheological Behavior Relevant to Extrusion Foaming

Richard Gendron

CONTENTS

2.1   Introduction: The Role of Rheology in Foam Processing

Extrusion of thermolastic foams is largely controlled by the relatively complex rheological behavior of the polymer melt and physical foaming agent (PFA) mixture [1]. The complexity arises first from the moderate to large plasticization due to the adequate dissolution of the small molecules of the foaming agent. They act as a diluent with respect to the polymeric macromolecules, which forms a single phase (zone #1, Figure 2.1). This solution should remain in that state until the nucleation step occurs at the die exit, as shown in Figure 2.1. This is usually accomplished by maintaining a pressure that is sufficiently high inside the extruder — i.e., above the solubility pressure for such conditions of temperature and PFA concentration — thereby preventing any phase separation. In the extruder, the deformations occur mainly in a shearing mode. Elongational and shear flows coexist in the die, where the PFA should still be kept dissolved in the polymer melt.

Knowledge of the viscosity decrease with respect to the PFA content is important, but obviously measurement cannot be performed using conventional rheometrical tools that operate at atmospheric pressure. A closed pressurized rheometer is thus necessary, which may require modification of an existing capillary rheometer [2] or implementation of an on-line rheometer on the foam extrusion line [3,4].

Nucleation occurs outside the shaping die (zone #2) due to the large pressure drop experienced at the die lips. Since the internal gas pressure inside a cell, formed by the blowing agent that has phase-separated from the molten polymer, exceeds the atmospheric pressure which now prevails for the melt, cell growth occurs, controlled by the extensional behavior of the resin matrix that forms the struts and walls of the cells.

Stabilization of the evolving cell structure (zone #3, Figure 2.1) can be controlled by various factors. For instance, depletion of PFA in the gas-swollen melt can induce a rapid viscosity increase if the processing temperature is close to the glass transition temperature of the neat polymer. Viscosity rise can also be induced in some other cases by crystallization,

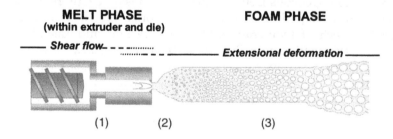

FIGURE 2.1
Schematic of the extrusion foam process. (1) A single-phase, gas-laden melt is pumped toward die exit. (2) Nucleation occurs outside the die due to the large pressure drop. (3) Cells expand and the structure stabilizes as the polymer is cooling.

but relying on this sole effect would make the processing window very narrow. Foaming linear polypropylene (PP) was known to be very difficult and possible only in a temperature range of a few degrees. Development of a technology that induces long chain branching in PP in the mid 1980s eased the foaming process for this resin. This chain structure is responsible of the abrupt rise in the transient viscosity response, which enables the cell structure to remain stable until crystallization occurs. The influence of the rate of crystallization on the foam processability has been highlighted by Constant [5].

Although the importance of rheology to foam processing is now being widely recognized, relevant scientific literature devoted to foaming considerations remains relatively scarce. This can be easily explained by the technical difficulties that arise when one tries to reproduce the typical conditions that prevail during the extrusion foaming process. First, commercial rheometers are not adapted for gas-ladden melts, where the single-phase state can only be maintained through adequate pressurization. Second, even though instruments can be modified to give access to the rheological behavior under shear deformations, availability of adequate reliable techniques is still lacking when it comes to extensional rheology. Even if very few commercial instruments are available, experimental difficulties are still associated with these methods [6], forcing investigators to look for alternative methods in order to assess the foaming performance of given resins.

One method that is being widely employed refers to the "melt strength" of the resin, which reports typically the maximum tensile strength measured during the continuous drawing of the extrudate from a capillary die. Due to its simplicity this technique is very attractive. However, it provides little information on the transient viscosity response that controls the cell growth phase. Moreover, even though this technique is based on the onset of rupture of the melt under elongational deformation, it does not provide relevant details on the mechanism of this rupture. In fact, this last topic has been practically ignored in the literature, with only a small number of publications devoted to this critical aspect.

In this chapter, the basic rheological mechanisms that prevail during the extrusion foaming process will be reviewed, with emphasis on the extensional behavior of relevant polymers. Particular attention will also be given to the conditions that provide a homogeneous deformation or, in the contrary, that lead to rupture and open cell morphology.

2.2 Basic Rheological Concepts

It is not the intent of this chapter to provide an extensive description of the rheology of polymer melts, but rather to give to the reader a basic understanding of some rheological concepts that will be related to the foaming process in the following sections. Some excellent books provide a more complete understanding of the rheological behavior of polymers, its relationship to the molecular structure of the macromolecules [7–9], and the various techniques that can be used to probe this rheology [10]. Readings can be further extended to more classical textbooks on viscoelasticity [11].

2.2.1 Dynamic Rheology

In dynamic oscillatory testing, the viscous and the elastic moduli, G'' and G' respectively, are evaluated as a function of frequency (Figure 2.2[a]). The strain that is imposed on the sample is usually kept at a relatively low level, such that the response is said to remain in the linear viscoelastic domain (with the same modulus irrespective of the level of strain). The loss and storage moduli can be combined to yield the complex viscosity η^*:

$$\eta^* = ((G'^{\,2} + G''^{\,2})^{1/2})/\omega \qquad (2.1)$$

The dynamic viscosity η' can also be calculated from the viscous component of the dynamic responses:

$$\eta' = G''/\omega \qquad (2.2)$$

The viscosity prevailing at low frequencies follows a Newtonian behavior (constant viscosity irrespective of the rate of deformation), and the value for this plateau viscosity is referred to as the zero-shear viscosity η_0 (Figure 2.2[b]).

Dynamic viscosity testing is a convenient way to have access rapidly and with little effort to a wide range of practical information. A fair estimate of the shear behavior is provided with the Cox-Merz rule, without undergoing tedious capillary rheometry measurements:

$$\eta(\gamma) = \eta^*(\omega)\Big|_{\omega=\gamma} \qquad (2.3)$$

Moreover, it is possible to derive information on the molecular structure of the resin using specific correlations, which is highly convenient for polyolefins that otherwise would require laborious high temperature gas-phase chromatography (GPC) for the generation of similar knowledge. This procedure is based on the intersection of G' and G'' curves, which yields the crossover modulus G_C and the crossover frequency ω_C [12]. These crossover quantities are respectively correlated to molecular weight distribution (MWD) and the weight-average molecular weight M_w. The reciprocal of the crossover modulus gives the polydispersity index (PI). However, as demonstrated by Bernreitner et al. [13], for various types of PP, the empirical correlations found are not universal, and should be restricted to families of polymers with similar structures.

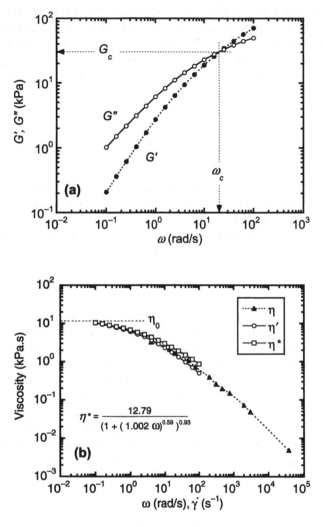

FIGURE 2.2
Dynamic and steady-state rheological behavior of a linear PP at 220°C. (a) Viscous (G'') and elastic (G') moduli, with location of crossover modulus G_C and crossover frequency ω_C. (b) Complex (η^*) and dynamic (η') viscosities, compared with steady-state viscosity η. The complex viscosity was fitted using Equation 2.4.

The complex viscosity η^* can be fitted using a modified Carreau equation (Equation 2.4), which uses four parameters: the zero-shear viscosity η_0, the mean relaxation time τ, and m_1 and m_2, the exponents related to the degree of curvature and the slope in the high frequency (or high rate) power-law region.

$$\eta^* = \frac{\eta_0}{(1+(\tau\omega)^{m_1})^{m_2}} \qquad (2.4)$$

Another valuable result that can be extracted from the dynamic measurements is the steady-state compliance J_e^0 [14], which is related to the elastic component of the deformation. It can be associated with the amount of strain recovery or recoil that occurs after the stress is removed. The steady-state compliance can be derived from the elastic modulus in the low frequency range using the following equation:

$$J_e^0 = \frac{G'}{\eta_0^2 \omega^2} \tag{2.5}$$

This value, related to the elastic response associated with long relaxation times, should also be sensitive to the degree of elasticity required during the cell growth and stabilization stage.

A similar rheological parameter is the loss tangent, which simply relates the relative magnitude of the two moduli:

$$\tan \delta = G''/G' \tag{2.6}$$

The way viscosity changes with temperature seriously affects the foamability of a resin, or at least the width of its processing window in terms of efficient temperature range. The dependency of viscosity on temperature can follow two distinct relationships, given that the polymer is amorphous and near its transition into a glassy state, or that the polymer will undergo partial crystallization in a close temperature range (still being very far from its glass transition temperature). The former case is the one that corresponds to polystyrene, poly(methyl methacrylate), and polycarbonate. For temperatures not greater than 100°C above the glass transition temperature T_g, the temperature-dependent viscosity can be described using the well-known Williams-Landel-Ferry (WLF) equation [15]:

$$\log \eta/\eta_g = -\frac{c_1 (T - T_g)}{c_2 + T - T_g} \tag{2.7}$$

with c_1 and c_2 being constants that reflect the temperature-dependent free-volume fraction.

Otherwise, for semicrystalline polymers like polyolefins, the Arrhenius-type equation is recommended, with a constant value for the energy of activation E_a (in this equation, the temperature is expressed in Kelvin):

$$\eta = A + \exp (E_a/RT) \tag{2.8}$$

This equation can also be used for amorphous polymers when temperatures are greater than $T_g + 100$. The viscosity will be more sensitive to temperature variation with increasing E_a value. In general, the viscosity of a linear

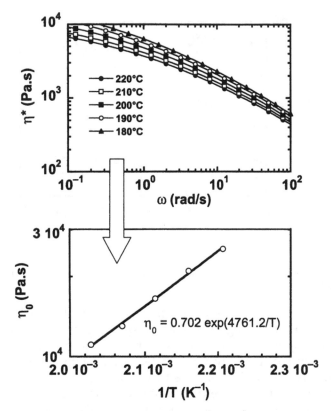

FIGURE 2.3
Computation of the energy of activation for a linear PP. (Top) Complex viscosity curves at different temperatures, from which are extrapolated the η_0 values (Equation 2.4). (Bottom) Fit using Equation 2.8, with the slope being equal to E_a/R (with R = 8.314 J/(mol K), and E_a in J/mol). E_a is thus equal to 4761.2 × 8.314 = 39585 J/mol.

polyolefin is less temperature-dependent than its branched counterpart; nevertheless, polystyrene and other amorphous polymers exhibit much greater dependency (for example, 25 kJ/kg mol for high-density polyethylene [HDPE] vs. 100–150 kJ/kg mol for polystyrene [PS], under standard processing temperatures). An illustration of the use of Equation 2.8 for the determination of the energy of activation is given in Figure 2.3.

2.2.2 Shear Rheology

Steady-state shear viscosity determination can be carried out using either a cone-and-plate rheometer or a capillary viscosimeter. The second method, although more time-consuming, is usually preferred since the shear rate range that can be investigated is much closer to that experienced for several processes such as extrusion and injection molding.

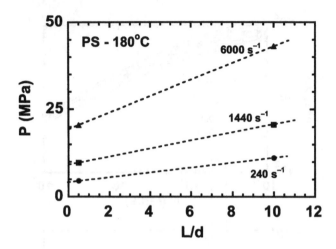

FIGURE 2.4
Bagley plot for a polystyrene resin tested at 180°C at three different shear rates, using two capillaries with different L/d.

The capillary method consists in pumping the molten polymer at controlled piston speeds through a capillary of known geometry (capillary diameter and die length). The shear stress at the wall is related to the pressure drop in the capillary, for a given flow rate. However, the measured pressure drop also incorporates a loss of pressure due to entrance effects (extensional flow). Since the total pressure drop is the sum of pressure loss due to the entrance converging flow (independent of the capillary length) and the pressure loss due the shear stress in the capillary at the wall (proportional to the capillary length), two capillaries having the same diameter but with different lengths are at least required to extract the respective contributions of the shear stress and the entrance effect. Overall pressure drops, obtained with different capillaries but at the same shear rate, can be plotted as a function of the *L/d* ratio (commonly known as the Bagley plot; see Figure 2.4). Entrance loss and shear stress (σ_{12}) will be obtained through a linear fit:

$$P = P_0 + k_1 \, (L/d) \qquad (2.9)$$

where P_0 corresponds to the entrance losses, and $k_1 = 4\sigma_{12}$.

The apparent shear rate at the capillary wall γ_{app} is given by:

$$\gamma_{app} = \frac{32\,Q}{\pi d_c^{\,3}} = \frac{8v_p d_p^{\,2}}{d_c^{\,3}} \qquad (2.10)$$

where Q is the polymer flow rate, d_c is the capillary diameter, v_p is the velocity of the piston, and d_p is the piston diameter.

The term *apparent shear rate* is used here because Equation 2.10 is valid only for a Newtonian fluid. If the fluid behaves as a shear-thinning, power-law fluid, then the shear rate should be corrected (Rabinowitch correction) with respect to the power-law index n ($n = \delta \log \sigma_{12} / \delta \log \gamma_{app}$):

$$\gamma = \frac{3n+1}{4n} \gamma_{app} \qquad (2.11)$$

An example of steady-state viscosity, as determined from capillary measurements, is displayed in Figure 2.2(b).

2.2.3 Extensional Rheology

The investigation of elongational behavior of polymer melts originated around 1970. These pioneering researches were conducted to elucidate the reason why two polymers exhibiting similar shear behavior did not perform identically in processes where extensional flow fields were dominant. Thirty-five years later, access to reliable extensional rheometers is still limited. Most of the rheometers used during that period were homemade, and very few have been commercialized, even over a limited period of time [16,17]. However, extensional rheometry has always suffered from the complexity of the experimental procedure required to generate reliable results under well-controlled conditions [18]. Ongoing developments are still underway to fill the gap between research and industrial requirements and to ease the job of generating reliable data [19].

All this may explain the popularity of the haul-off experiment that leads to the melt strength measurement. This technique is often called the Rheotens experiment, named after the tensile testing instrument developed by Meissner [20]. The melt strength is defined as the maximum tensile stress, or force at break, measured during the continuous drawing of an extrudate from a die, with a gradual increase of the haul-off speed. The same experiment also provides an estimate of the extensibility of the polymer melt, from the maximum value achieved in the rotation speed of the haul-off device at the point of rupture. From the standard method, however, no information on the transient viscosity response can be extracted, this method relying essentially on the ultimate rupture conditions. Calculation of elongational viscosity from the Rheotens curves is nevertheless possible, provided optical monitoring of the extrudate diameter [21]. Thus, although widely used, this technique can only provide qualitative estimates of the extensional rheology, the same way its rheological counterpart, the melt index (MI) or melt flow index (MFI), only partially reflects shear viscosity behavior.

Elongational or extensional (stretching) deformation of polymer controls many important processes, such as fiber spinning, film blowing, thermoforming, and extrusion foaming (bubble growth in cellular extrudates). And for such processes that are driven essentially by extensional flows, the

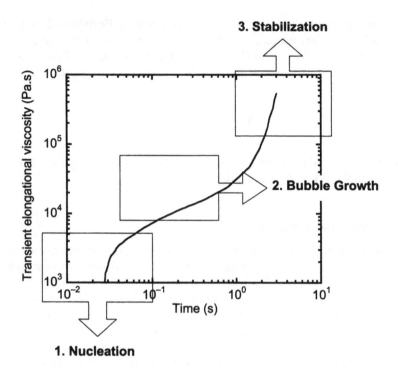

FIGURE 2.5
Zone 1: Initial slope of the stress growth function, which correlates to the polydispersity index.
Zone 2: Linear viscoelastic behavior. Zone 3: Nonlinear viscoelastic behavior highlighted by
the presence of strain hardening.

deformations take place over a limited period of time, so the flow does not
have time to get fully developed until a steady-state level (i.e., stable stress
over time or strain) is reached. For these processes, including foaming,
transient elongational flow properties are more relevant than steady-state
viscosities. Figure 2.5 illustrates a typical transient elongational viscosity
response, obtained for a given strain rate, with indication of the critical stages
occurring during foaming. Obviously this illustration is not fully accurate,
since the rate of deformation prevailing during cell growth should be variable, from a high to a low value. We see nevertheless from Figure 2.5 that
the transient viscosity η^+_E evolves with time t when submitted to a constant
deformation or strain rate ε, or, stated differently, with the amplitude of the
increasing deformation (Hencky strain $\varepsilon = \varepsilon\, t$).

During the bubble growth, the cell walls experience biaxial stretching. The
same mode of deformation also prevails in polymer processes such as
thermoforming, film blowing, and blow molding. But biaxial stretching is
only one of the three modes of extensional flow, the other two being uniaxial
(as in fiber spinning) and planar (wire coating). Extensional viscosity is
denoted η_U, η_P, and η_B for uniaxial, planar, and biaxial deformation, respectively (Table 2.1).

TABLE 2.1

Types of Elongational Deformation Related to Polymer Processing

Deformation	Viscosity	Schematic	Process Examples	Trouton Ratio T_R	Strain Hardening
Uniaxial	η_U		Fiber spinning	3	Maximum
Planar	η_P		Rolling, wire coating	4	Moderate
Biaxial	η_B		Film blowing, foaming, blow molding	6	Minimum

A simple relationship, obtained by Trouton in 1906, links elongational and shear viscosity for Newtonian liquids:

$$\eta_E = 3\,\eta \tag{2.12}$$

with the shear viscosity denoted simply by η. The concept of a Trouton law has been extended to viscoelastic fluids and other flows with the Trouton ratio T_R, defined as follows:

$$T_R = \frac{\eta_E(\varepsilon)}{\eta(\gamma)} \tag{2.13}$$

Trouton ratios T_R are 3, 4, and 6 for uniaxial, planar, and biaxial flows respectively, but these numbers hold only in the limit of vanishing low deformation rates, as displayed in Figure 2.6:

$$\lim_{\varepsilon \to 0} (\eta_E/\eta_0) = T_R \quad \text{where} \quad \eta_0 = \lim_{\gamma \to 0} (\eta) \tag{2.14}$$

Long-chain branching (LCB) in polyolefins is often associated with the occurrence of "strain hardening" (SH) in the transient viscosity curve. SH is generally related to the inability of the macromolecules to disentangle quickly enough to follow the exponential deformation. This helps to generate a homogeneous and stable cell structure by preventing cell rupture in the highly deformed cell walls.

The onset of the strain-hardening feature, i.e., the deviation from the linear viscoelastic growth curve, occurs at a constant value of the strain ε_{SH}, which may vary from one polymer to the other, but remains approximately

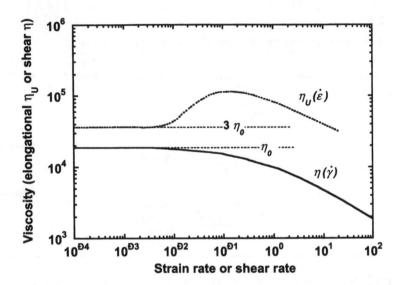

FIGURE 2.6
Schematic comparison between elongation (uniaxial) η_U and shear η viscosity responses for LDPE.

independent of the elongational strain rate, as shown in Figure 2.7(a) [20]. Moreover, as illustrated in Figure 2.7(b), for the transient elongational viscosity expressed as a function of the strain, all curves merge in the SH zone, which supports the idea of a rubber-like behavior driven by a near perfect elastic behavior, with no incidence of the strain rate on the viscosity-strain dependency. While the viscous component is controlled by the relaxation times of the macromolecules, the stored elastic part would be simply driven by the deformation.

While LCB is unambiguously at the source of SH, other molecular parameters, such as polydispersity (or MWD) and bimodality of the MWD (with presence of high molecular weight components) can also contribute to the occurrence of an apparent SH effect [22].

The magnitude of SH, for a given polymer, is dependent on the mode of extensional deformation. While planar deformation induces moderate SH, maximum SH is observed under uniaxial flow and minimum SH for biaxial stretching [23]. Biaxial flow causes the macromolecules to be pulled away following orthogonal directions, which induces disentanglement and thus leads to SH reduction.

The SH can be quantitatively reported with respect to the linear viscoelastic response, by comparing the true response to the expected one:

$$\lambda_{SH}(\varepsilon, t) \equiv \eta^+_{E,obs} / \eta^+_{E,linear} \qquad (2.15)$$

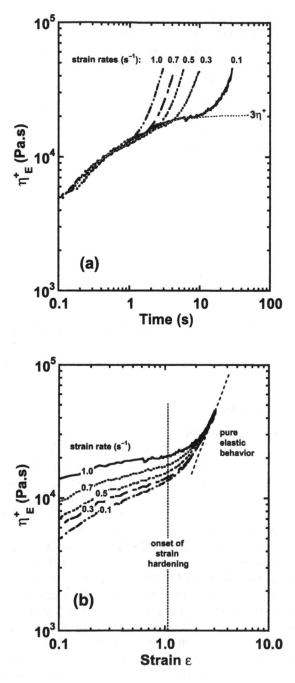

FIGURE 2.7
Different representations of the strain hardening behavior for a branched PP, characterized at various strain rates: (a) As a function of time (superposition of the linear viscoelastic portion) and (b) as a function of strain (merge of the elastic behavior).

Other methods can provide valuable information about elongational behavior. While the storage modulus G' and the first normal stress difference N_1 are considered direct measures of melt elasticity, the latter can also be estimated from indirect measurements such as the entrance pressure drop P_0 (the so-called Bagley correction, Equation 2.9), and extrudate swell B.

Elongational measurements can also be derived according to Cogswell's equation from the capillary entrance pressure drop [24]. The elongational stress σ_E is related to the pressure drop in the orifice die or to the Bagley correction P_0 through the following:

$$\sigma_E = \frac{3}{8}(n+1)\,P_0 \tag{2.16}$$

The strain rate ε is given as follows:

$$\varepsilon = \frac{4\sigma_{12}\gamma}{3\,(n+1)P_0} \quad \text{or} \quad \eta\gamma = 2\eta_E\varepsilon \tag{2.17}$$

where n is the power-law slope of shear stress vs. the shear rate in capillary flow, γ is the shear rate in a capillary of equivalent diameter, σ_{12} is the shear stress and η and η_E, the shear and extensional viscosity, respectively.

2.2.4 Rheological Models

The development of a constitutive equation relating stress to deformation history has been complicated due to the rheological effects of LCB. The single integral model of the Bernstein–Kearsley–Zapas (BKZ) type has enjoyed popularity in the past due mainly to its robustness and its flexibility in fitting the rheological uniaxial extension and shear responses through kernel functions [25]. This model takes into account that time and deformation, as experimentally observed, are separable. A first function, called memory function, describes the linear viscoelastic effects through discrete spectra of relaxation modulus and relaxation time, while a second function, the damping function, assumes that damping for both shear and extensional flows follows an exponential form. These functions are finally convoluted in the time domain with a strain function (Finger and Cauchy tensors).

Unfortunately, this model still fails for certain applications, such as planar extension of low-density polyethylene (LDPE) [26]. The recent development of a molecular constitutive theory based on an idealized branched melt, called the pom-pom model, has gained adepts due to its simplicity, and matches adequately the unusual branched-structure flow behavior.

The molecular architecture chosen for this model can be schematically described as an H-polymer structure with two branch points, between which is the "backbone" of molecular weight M_b; each branch point possesses q

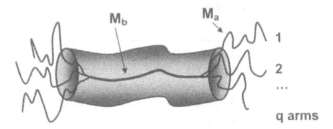

FIGURE 2.8
Schematic representation of an H-polymer structure, with the backbone of the molecule inserted into the Doi–Edwards tube, and with q branches (or arms) on each side of the backbone.

arms of molecular weight M_a (see Figure 2.8). The tube model developed by Doi and Edwards [27] is applied here to the special case of branched polymers, in which the dynamics of entangled polymers consist of diffusion/reptation and stretch/retraction. The unique properties of branched polymer melts are thus associated with the restriction imposed to the retraction of the segment between branch points, compared to the more mobile chains (or arms) with free ends. This obviously imposes a strong separation of relaxation timescales between the backbone, especially for the segments near the branch points and the arms. Relaxation of these arms occurs first, and their timescale is relatively small compared to the constrained retraction of the backbone.

Nevertheless, very small deformation rates allow for complete relaxation of the branched structure network, with no sign of SH [20]. However, under faster rates, uncompleted relaxation will mimic a crosslinked network, with a significant elastic behavior and SH response.

2.3 Review of Extensional Behavior of Various Foamable Polymers

2.3.1 Polystyrene

One of the very first thorough investigations of the effect of the molecular weight distribution on the elongation flow of PS was conducted by Münstedt in 1980 on four different PS resins [22]. Deviations from the linear viscoelastic behavior were observed for the resin having bimodality in its molecular weight distribution, with the presence of strong SH transient response. Significant SH was also present for resins widely polydispersed, or having a content of very high molecular weight. It was later proposed that relaxation times above 1000 s, provided by the high molecular weight fraction, could be responsible of the SH behavior. This was validated with a high molecular weight fraction as low as 0.8 wt% [28].

Impact of the relaxation times on the SH feature will be illustrated in the next example. The transient viscosity curves of a commercial polystyrene are depicted in Figure 2.9(a) for temperatures ranging between 150 and 185°C, and for strain rates between 0.1 and 1.0 s^{-1}. SH gradually takes place as the temperature is lowered and as the deformation rate is increased. As reported by Münstedt [29], the time-temperature superposition is applicable to elongational flow of PS despite its nonlinear viscoelastic signature. Figure 2.9(b) was obtained through curve-shifting, based on shift coefficient a_T computed using dynamic rheological results performed over the same temperature range. It should be noted that the flow energy of activation of amorphous polymer is not a constant value as for polyolefins, and that the viscosity results should follow the WLF equation (Equation 2.7). Overlapping of corresponding curves was excellent, even in the nonlinear portions. So Figure 2.9(b) provides a good illustration of the rising of the SH function as the strain rate is increased.

Molecular relaxation of both mono- and polydispersed PS resins has been studied during and after a step-strain uniaxial deformation at temperatures typically in the 100–160°C range [30]. Interestingly, this temperature range is the one prevailing for PS foam extrusion, made possible by the presence of plasticizing foaming agent. Relaxation measurements were then deconvoluted into three different mean relaxation times. The shortest relaxation time τ_1 was in the order of seconds, and was independent of molecular weight. The second, τ_2, was typically in the 1-minute range while the longest, τ_3, was typically a few thousand seconds. The last two relaxation times were respectively associated with partial relaxation (τ_2) and recoil (τ_3) of the chain end, and they scaled with the molecular weight. These different timescales suggest a behavior similar to that of branched molecular architectures as represented by the pom-pom model (Section 2.2.4). The longest relaxation times observed for PS would explain the strong elastic behavior under low temperature and high strain rate conditions.

The impact of a star-branched structure on rheology and foaming has been investigated. In this study, star-branched PS have been compared to a group of linear PS of different molecular weights. The complex viscosity curves of these polymers have been fitted with Equation 2.4 to provide both zero-shear viscosity and mean relaxation time data. Estimates of the steady-state shear compliance (Equation 2.5) and the PI were also derived from the complex rheologial data. As illustrated in Figure 2.10(a), the mean relaxation time is simply proportional to the zero-shear viscosity (molecular weight) and no specific information can be retrieved for the star-branched PS. However, the steady-state shear compliance is sensitive to the molecular weight distribution (PI) and is strongly dependent on the branched structure of the PS, as shown in Figure 2.10(b). Impact of the enhanced elasticity is confirmed through uniaxial elongational rheology tests, where the level of SH is slightly magnified with the presence of a branched structure (Figure 2.11). On the same figure several results are also plotted from the linear PS of different molecular weights. Once scaled according to their zero-shear viscosity η_0, a

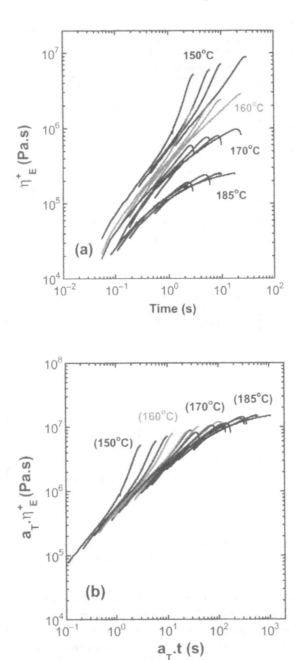

FIGURE 2.9
Transient elongational viscosity response obtained for various strain rates, at four different temperatures. (a) Raw results. (b) Shifted curves according to the time–temperature super-position principle.

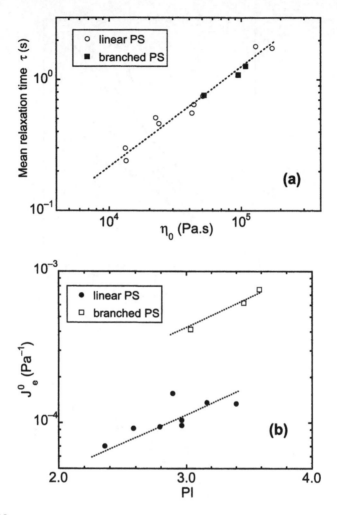

FIGURE 2.10
(a) Mean relaxation time as a function of the zero-shear viscosity. (b) Steady-state compliance as a function of the polydispersity index (PI) for various polystyrene resins with different molecular structures.

similar portrait to that obtained with one PS at several temperatures (Figure 2.9) is finally obtained. However, contrary to polyolefins where SH is crucial for the foaming process, the star-branched PS behave very similarly during the foaming process, with a slight improvement in the cellular structure. This could be attributed by the different stretching modes, uniaxial for the characterization and biaxial during foaming.

However, processability of PS relies mainly on its high level of plasticization achieved in presence of the foaming agent. As this one phase separates from the polymer matrix, gradual depletion of the diluent makes the viscosity of the PS rise abruptly. As will be reported in Section 2.4.1, viscosity changes associated with the PFA can be a hundredfold, which would make

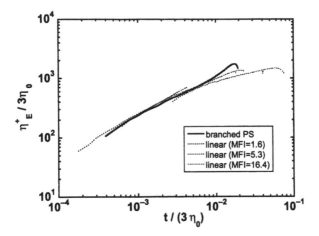

FIGURE 2.11
Elongational viscosity responses (normalized with respect to the zero-shear viscosity) for PS resins of different molecular structure; strain rate was set to 1.0 s^{-1}, and temperature to 180°C.

the transient viscosity rise at a rate similar to that obtained for a polymer exhibiting SH, as shown in Figure 2.12. This example reflects the transient viscosity behavior resulting from the successive viscosities observed at different times and excerpted from the transient curves shifted according to their plasticization level. This depletion-induced viscosity rise would make the relatively small SH effect of linear or star-branched PS negligible, thus less relevant in terms of its impact on processability.

2.3.2 Polyolefins

One of the very first published results on transient elongational viscosity was based on LDPE [20], with tests performed using strain rates varying between 0.001 and 1.0 s^{-1}. Interestingly, while the very low rates that have been investigated induced little or no SH, the magnitude of the SH increased with the strain rate, with maximum SH in the 0.1–1.0 s^{-1} range. The same publication also presented results on recoverable tensile strain ε_R. The relative amount of ε_R respective to the overall strain increased with the strain rate, with less than 25% for low rates below 0.01 s^{-1}, but almost 100% of the deformation being recoverable (purely elastic behavior) as the strain rate was increased above 1.0 s^{-1}.

Before the technological breakthrough of metallocene catalysts, polyethylene foaming was possible only in the presence of a significant fraction of LDPE. The presence of long-chain branchings inherent in LDPEs was mainly due to homogeneous deformation across the cellular structure as well as stabilization, through the rheological feature known as strain hardening. Magnitude of SH has been unambiguously correlated with the content of LCB [31]. Moreover, not only the number of branch points but also the

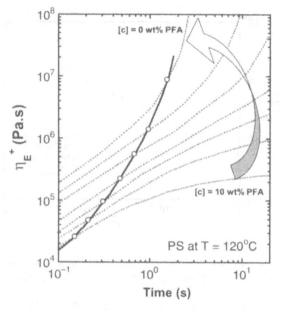

FIGURE 2.12
Simple model illustrating the continuous increase of the viscosity as the PS melt is gradually depleted from the dissolved foaming agent that was formerly acting as a plasticizer. Assumptions were made about the viscosity response and the depletion rate.

architecture of the branching, as well as the branch length distribution, impacts the rheological behavior and incidentally the degree of SH. For example, LDPEs manufactured from a tubular technology exhibit a "comb" type branching that is associated with SH of lower amplitude than that of LDPEs produced from the vessel technology, which yields a complex "tree" type branch structure, with an increased number of branch points (Figure 2.13) [32]. The benefit resulting from this branched structure can be simply stated as an increase in the melt strength while the melt index is maintained constant. Linear PE usually does not show any SH. However, some exceptions have been reported, for example linear low-density polyethylene (LLDPE) having a very broad molecular weight distribution with $M_z/M_n > 40$ [33].

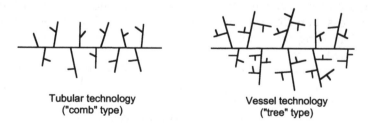

Tubular technology
("comb" type)

Vessel technology
("tree" type)

FIGURE 2.13
Schematic representation of the different types of branching structure obtained for different LDPEs as a function of their respective polymerization technology.

Presence of LCB in metallocene-type polyethylenes (mPEs) can be probed in the linear viscoelastic (LVE) domain, with increase of the storage modulus G' at low frequency and appearance of a plateau in the loss angle plot vs. frequency, with loss angle defined as the arctan of G''/G' [34]. These effects imply the occurrence of a new relaxation regime.

The flow energy of activation E_a is a mean value that is either obtained from the overall relaxation time spectrum shift, or from scaling of the zero-shear viscosity plateaus (Figure 2.3). Low values are associated with linear PE (27 to 30 kJ/mol, for conventional HDPE), with slightly higher values for LLDPE (30 to 35 kJ/mol), and much higher ones for branched LDPE (above 40 kJ/mol), a higher E_a value meaning that the viscosity is more sensitive to temperature variations. Branched polymers are termed thermorheological complexes, as compared to linear polymers, which exhibit much simple rheological behavior. As reviewed by Wood-Adams and Costeux [35], branch length and structure directly modify the apparent zero-shear activation energy. This implies that the long relaxation times associated with LCB are more temperature-sensitive. It was also reported that combination of LBC and short-chain branching (SCB) as found in vessel LDPE is synergistic and yields very high activation energy (58 kJ/mol) [35].

Unfortunately, this complex entanglement network associated with branched polymers, required for extensional flow property enhancement, is shear-sensitive, as illustrated in Figure 2.14. It has been reported that the shear history, a function of shear rate and duration of shear, significantly but reversibly lowers the melt strength [32]. The melt can recover its original properties through exposure to annealing, owing to micro-Brownian motion of the branches. Since the affected temporary couplings are those associated with very long relaxation times, it is not surprising to end up with required annealing times on the order of 10 to 30 minutes! Obviously, the more complex the branch network, such as that of vessel-type LDPE, the longer the annealing time, as demonstrated by Yamaguchi and Takahashi [32]. Their research also showed that the largest difference in SH behavior between vessel LDPE and tubular LDPE was observed under uniaxial elongational flow; both viscosity responses were much alike when mesasured under biaxial extensional deformation, with a significant decrease of the SH magnitude. This last observation is in agreement with several other works that rank the SH as a function of type of flow: uniaxial > planar > biaxial.

Since the processing has a strong impact on the shear history of the melt, as well as its remaining extensional required features, the extrusion foam processes, especially those involving twin-screw extruders, which can be very shear-efficient, must be finely tuned in order to minimize the possible shear-disentanglement effect during the melting, PFA dissolution, and pumping stages.

Correlation between magnitude of SH and the foamability of dynamically vulcanized thermoplastic elastomers (TPEs) based on blends of PP and ethylene–propylene–diene (EPDM) terpolymer was attempted for different systems in which the linear PP was partially substituted with branched PP, in

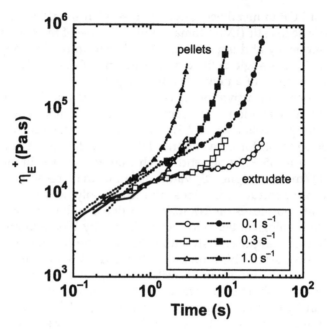

FIGURE 2.14
Differences in the transient elongational behavior for a branched-type PP before ("pellet") and after ("extrudate") extrusion, showing the changes experienced due to the disentanglement of the branched network.

order to enhance the melt strength [36]. With foaming performed with water at a concentration of 2.0 wt%, slight improvement in the density reduction, from 0.45 to 0.36 g/cm³, was obtained, switching from for the linear PP formulation to the PP blend that exhibited only moderate SH. Increasing the magnitude of SH through either an increase of the branched PP (bPP) fraction or the type of bPP surprisingly yielded poorer foam results, with a mean density of 0.64 g/cm³. This case illustrates well that while the extensional thickening induced by branched polymer can prevent cell coalescence and cellular structure stability, it can also be detrimental to the expansion, as it hinders bubble growth. Adequate balance between elastic and viscous contributions must therefore be sought, as detailed in Section 2.3.6.

2.3.3 PVC

The rheology of poly(vinyl chloride) (PVC), both for shear and extension flows, is rather complex due to its particulate structure, which evolves with shear and temperature. For example, unplasticized PVC (UPVC) will exhibit at a low temperature (but still above the glass transition temperature [$T_g \approx 80°C$]) a behavior similar to that of a lightly crosslinked matrix filled with tiny PVC particles, composed of crystallites and amorphous PVC parts (Figure 2.15). This stage is referred to as gelation, and exhibits maximum

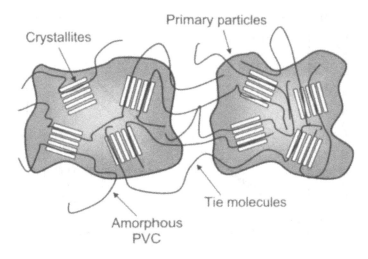

FIGURE 2.15
Schematic of PVC particles partially melted, with remaining crystallites inside the particles, and tie molecules between the particles.

viscoelastic properties at T_x = 180–185°C ($G' > G''$), this temperature being referred to as the fusion temperature. If the temperature is raised further, fusion of the remaining PVC primary particles will take place. The polymeric system behavior will be simply that of a filled polymer melt, until all PVC particles have disappeared. Above 210°C (T_m), the PVC is completely molten (the melting of crystallites is completed) and typical polymer melt flow curves are finally obtained, based on flow and relaxation of individual but still entangled macromolecules ($G'' > G'$).

If a plasticizer is used, the solvent molecules will be dissolved into the amorphous regions. The addition of a plasticizer (or a physical foaming agent acting as a plasticizer) will affect the three critical temperatures, T_g, T_x, and T_m, but will lower the glass-transition temperature [37] to a greater degree than the fusion and melting temperature [38].

Typical transient extensional viscosity curves of PVC, displayed as a function of time for various deformation rates, are illustrated in Figure 2.16. At the testing temperature of 190°C, which is below the melting temperature of the crystallites, the curves exhibit the SH feature at a Hencky strain of approximately 1. This behavior could be associated in this case with the presence of chain crosslinks induced by the remaining crystallites. Moreover, the strain-hardening effect vanishes when the temperature exceeds T_m, which supports the role of the crystallites in the chain networking mechanism.

It has been reported that a convenient way to enhance the extent of strainability of the melt relies on the addition of acrylic processing aids [40]. As shown in Figure 2.17, increasing the level of acrylic increases both the strain (extensibility) and the stress (resistance to rupture), with the rate of increase being very sensitive to the processing temperature. The role of the acrylic processing aid in postponing the tensile rupture was tentatively explained

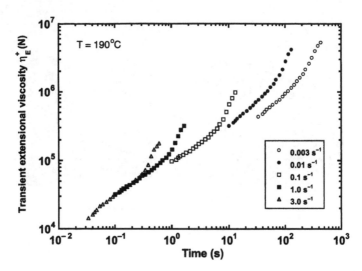

FIGURE 2.16
Transient extensional viscosity of PVC for various strain rates. (Adapted from Park, I.K. and Riley, D.W., *J. Macromol. Sci.-Phys.*, B20, 277, 1981 [39].)

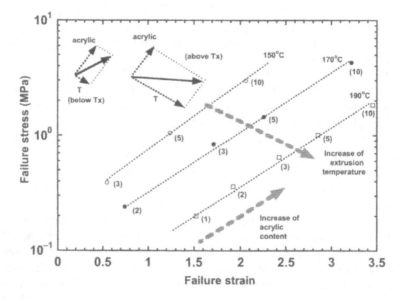

FIGURE 2.17
Stress and strain reached at rupture for a given PVC resin with various amounts of acrylic processing aid (indicated in parenthesis, in wt%), at different temperatures. (Adapted from Cogswell, F.N. et al., *Plast. & Rub.: Proc.*, 5, 128, 1980 [40].)

in terms of enhanced migration towards the potential cavitation sites, these latter inducing the rupture process.

Addition of acrylic processing aids is thus beneficial for PVC foaming, as illustrated in Figure 2.18 [41]. The lowest foam densities are in agreement

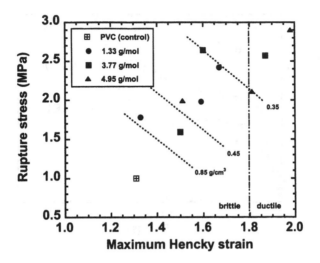

FIGURE 2.18
Impact of acrylic processing aids (with various molecular weights as identified) on the conditions experienced at rupture. (Adapted from Haworth, B. et al., *Plast. Rub. Comp. Process. Appl.*, 22, 159, 1994 [41].)

with the extensibility of the melt, as measured through the maximum Hencky strain achieved at rupture. The rupture mode associated with breakage of the PVC filament for strains roughly above 1.8 was typically a ductile mode, while brittle ruptures were encountered when breakage occurred at lower strains. Expansion appears to be limited by the elastic behavior of the deformation when the processing aid content is low, or its type less efficient. As deduced from elastic recovery data generated within the same study, increasing the effectiveness of the processing aid induces more permanent or irreversible deformation (viscous), which improves the stretchability of the melt, allowing for much lower foam densities.

2.3.4 PET

Most commercially available polyethylene terephthalate (PET) grades are not suitable for processes that involve stretching deformation because of the low viscosity of the resin and its well-known poor melt strength. Modification of the based resin in order to induce chain extension and especially LCB can be performed by reactive processing through combination of the PET end-groups with selected chemical modifiers. Branched structures are obviously at the source of the induced strain-hardening effect reported in Figure 2.19 for a high molecular weight PET (intrinsic viscosity of 1.02) modified with various contents of a terpolymer of ethylene, methyl acrylate, and glycidyl methacrylate (E-MA-GMA).

However, performing elongational viscosity measurements remains a formidable task, and other methods to assess the degree of branching and the

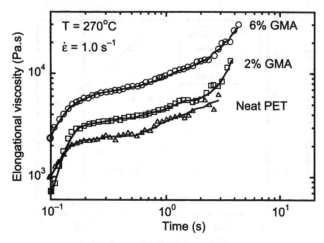

FIGURE 2.19
Extensional viscosity results for PET: neat resin (results on bottom) and chemically modified branched PET (top two curves), showing increased viscosity and strain hardening.

increase in elasticity have been reported. For instance, increase of the energy of activation, as determined from complex viscosity measurements, has been reported with increasing the amount of E-MA-GMA [42]. The energy of activation of the unmodified PET, having an intrinsic viscosity (IV) of 0.80 dL/g, was approximately 87 kJ/mol, and its dependence with the modifier was 2.3 (kJ/mol)/(wt% of E-MA-GMA).

Branching and crosslinking were also performed on PET using a tetrafunctional epoxy [43]. The reported increase of the IV was from 0.82 for the neat PET to 1.05 dL/g with an epoxy content of 0.25 wt%, and this was associated with a 25-fold increase of the apparent uniaxial elongational viscosity as measured using a haul-off system installed at the exit of the twin-screw extruder.

Most extruded viscoelastic polymers exhibit extrudate swell, with swell being defined as the ratio of the extrudate diameter to that of the capillary die. The swell can be related to elastic recovery, and it increases with higher stresses prevailing in the die [44]. The enhancement of elasticity through reactive processing of PET has also been assessed by the change in the swelling of the extrudate, as measured at a constant shear rate (270 s^{-1}) and constant temperature (260°C) [45]. Furthermore, these results were correlated with the foamability of the resin: unmodified PET exhibiting little or no swell (1.0–1.5) yielded poor foams, while modified or branched PET, with good performance during foam expansion, was associated with a higher swell (2.5–3.0). This study also indicated that the intrinsic viscosity could not be used as a good indicator for foamability. While base resins having IV in the 0.7–1.0 range were inappropriate for foaming, the best foams were still obtained with IV values of 0.87–0.95. The IV number, being a function of the hydrodynamic volume of the macromolecule, may correlate well with the molecular weight for linear chains, but fails to give appropriate information on the branched structure that provides elasticity.

2.3.5 Polycarbonate

Foaming of polycarbonate (PC) has been mainly conducted in the batch process, and only a few studies report extrusion foaming of this amorphous polymer. Surprisingly, while similar results from extrusion foaming were obtained for two different grades of PC, a linear vs. a branched one, the same grades yield significant differences in the cellular structure when processed in batch foaming, with the branched type providing a much higher cell density [46,47]. The difference in the results, batch vs. extrusion, could lie in the range of temperature involved for the two processes, with the much lower temperatures used for batch foaming resulting in longer relaxation times. This assumption is validated through the results reported by Lee et al. [48], that showed huge differences in the PC foam properties as the extrusion temperature was lowered, low temperatures being accessible because of the plasticization effect of the foaming agent, as commonly experienced with amorphous resins.

Because of its high glass transition temperature (150°C) and associated elevated processing temperatures, rheological testing under elongational deformation of the neat PC has been difficult to perform, because most of the techniques available require the immersion of the polymer sample into a bath of a suspending fluid (generally a silicon oil) which unfortunately cannot sustain the high temperatures involved with PC. Use of a cushion of nitrogen, as for the newly developed, but no longer manufactured RME (Rheometrics™ Elongational Rheometer for Melts), enables elongational testing of these resins [6]. This is illustrated in Figure 2.20 for the branched PC that exhibits SH responses of high magnitude. Impact of the branched structure is also put in evidence through the increase of the energy of activation (from 99.0 for linear to 113.0 kJ/mol for the branched resin) and also through the increase of the storage modulus G' at low frequencies, as illustrated with tan δ in Figure 2.21.

2.3.6 Defining Appropriate Rheological Behavior for Foaming

Because of the difficulties associated with the restricted access to reliable extensional data, it was found to be very attractive to evaluate the foamability of a resin quantitatively in terms of its relative elasticity, as probed through the dynamic rheological behavior.

Use of tan δ to assess the foamability of resins has been reported in the literature. In one case [49], the final foam product characteristics (foam density ρ and average cell size D) were related to the resin properties (tan δ

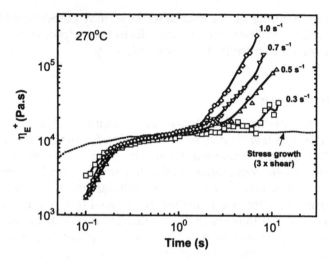

FIGURE 2.20
Apparent elongational viscosity measurements for a branched polycarbonate.

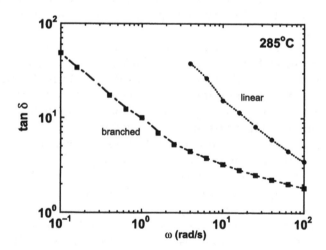

FIGURE 2.21
Differences in the loss tangent between linear and branched PC resins.

measured at 190°C and 1 rad/s). The requirement for the foamability factor F was also stated so as to yield closed cell polypropylene foams:

$$F = \rho D \, (\tan \delta)^{0.75} \leq 1.8 \qquad (2.18)$$

A slightly more complex, although qualitative, approach was attempted in a study based on the investigation of the foaming of low density poly(ethylene-co-octene) resins of different melt indices and various crosslinking agent levels, with formulations specially designed for the injection molding process [50]. It was concluded that the complex viscosity and the loss tangent were two rheological parameters that were complementary to each other, and both were required to define the foaming capability window that would yield minimum foam density and acceptable cell morphology. This can be schematically represented by plotting the loss tangent as a function of the complex viscosity, with these two values defined at $\omega = 10$ rad/s (Figure 2.22[a]). The effect of the molecular weight for the base resin on the relationship between elasticity and shear viscosity is given on this graph by a slope less steep than the one observed for the crosslinked structure induced by the addition of peroxide, indicating that the elasticity, associated with tan δ, is mainly induced by branching during crosslinking, and to a lesser degree by linear chain extension. The crosslinking follows a path for each base resin, and these paths or slopes are parallel. The crosslinking induces higher molecular weights but also implies a higher degree of long chain branching up to the gel point, which shifts the slope associated with crosslinking to the right.

An optimal processing window was then defined in terms of the aforementioned rheological criteria, as represented by the shaded area in Figure 2.22. Bubble growth would be controlled by the complex viscosity, while stabilization of the final morphology would be linked to the elastic properties as evaluated through the loss tangent.

2.3.7 Controlling the Strain-Hardening Feature

As seen in the previous sections, the branched structure of the polymer impacts significantly its nonlinear viscoelastic response during elongational flow. Bimodal molecular weight distribution would also introduce an SH-like response, although the mechanisms might be different. Since the SH signature is intimately related to enhanced processability during foaming, modification of existing resins in order to induce such a rheological characteristic, as for linear polymers, might render them suitable for foam processing.

Post-polymerization operations, such as grafting and crosslinking, can be achieved during the extrusion or molding process, prior to the foaming step. Starting with a polymer having a dominant linear structure, peroxide [50] or radiation [51] crosslinking methods induce numerous structure modifications, leading to molecular weight increase, the creation of LCBs and, ultimately, complete crosslinking resulting in increasing gel fractions. Final polymer structure, controlled by the extent of the reaction, depends on many variables, such as irradiation dose or peroxide content and starting polymer structure, as well as any other variables that may affect the reaction kinetics (temperature, additives, etc.). Figure 2.23 illustrates two features linked to the structural change of a low viscosity LLDPE modified by a peroxide-induced crosslinking reaction. First, the plateau associated with the zero-shear viscosity is shifted upward by almost two decades, indicating a tremendous increase in the molecular weight. Second, the presence of strong

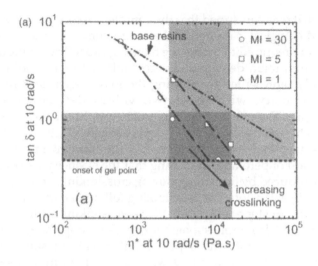

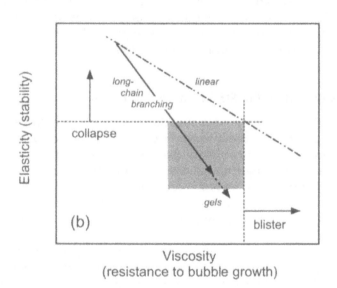

FIGURE 2.22
(a) Use of two rheological parameters, tan and complex viscosity, to define the optimal processing window, schematically represented in (b) according to the corresponding elasticity and viscosity characteristics. (From Gendron, R. and Vachon, C., *J. Cell. Plast.*, 39, 71, 2003 [50]. With permission.)

SH at every strain rate tested is obviously linked to multiple LCBs. The overall modified rheological response was such that foams of low density and acceptable cell structure have been produced [50].

But blending probably remains the easiest and most practical way to modify the rheological behavior and, for these reasons, is a common industrial

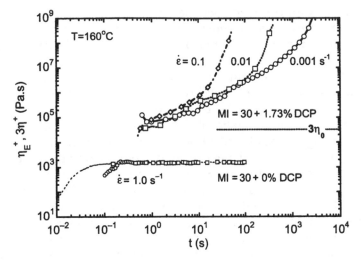

FIGURE 2.23
Extensional viscosity results for both noncrosslinked resins (bottom curve) and a crosslinked mixture (1.73 wt% of dicumyl peroxide [DCP], top three curves), for various strain rates, with associated stress growth correspondence. (From Gendron, R. and Vachon, C., *J. Cell. Plast.*, 39, 71, 2003 [50]. With permission.)

practice in the polyolefin foam industry, as reported in Chapter 3. For instance, increasing the amount of LDPE to LLDPE resulted in a gradual departure at a high strain rate from the linear viscoelastic response associated with the linear polyethylene [52]. Thus the SH increased linearly with a fraction of LDPE in that immiscible blend. Also, the same study reports that blending two LLDPEs, one standard and one having a very broad MWD, induced strong SH enhancement, even if nonadditivity was observed, with as low as 20 wt% of the broad MWD LLDPE. Targeting applications based on the extent of expansion is also possible by tuning the amount of the branched resin blended with its linear counterpart [53]. While high-density foams require only a small amount of branched resin, achievement of low-density foams may require extensional properties only possible through blends where the SH-responsible polymer is predominant.

Adding a small amount of ultra-high molecular weight (UHMW) polymer to a standard matrix also enhances the SH behavior, although this effect might be governed by the longer relaxation times associated with molecular weights above one million [28]. In order to be efficient, the added component must be miscible with the matrix; otherwise the rheological behavior remains that of the unchanged matrix [54]. For the systems investigated, either miscible or immiscible, the linear viscoelastic response remains unaffected.

Another route for enhancing the SH feature has been proposed by Yamaguchi et al. [55], based on the miscible blend of a crosslinked polymer having a low density of crosslinked points and a linear polymer. Drastic impact on the SH was observed with a composition of the gel-rich fraction as low as 3 wt%, while only moderate increase of the SH was induced with blends containing the soluble fraction obtained from the same peroxide-crosslinked HDPE. Similar observations were made on the miscible blends of isotactic PP and crosslinked ethylene-1–hexene–ethylidene norbornene (EHDM) terpolymer [56].

On the other hand, elimination of SH may result from the addition of organic or inorganic filler particles. Suspensions of polyisobutene with spherical monodispersed PS beads of various diameters have been investigated under uniaxial elongational deformation [57]. It was reported that either increasing the bead content or reducing the sphere size significantly diminishes the SH, and the latter can almost disappear with particles having diameters in the 180 nm range. Similar observations were made for glass-bead-charged HDPE systems, the polymer having a broad and bimodal molecular distribution [58]. The strongest negative impact on the SH has been reported for systems containing either talc or glass fibers, with a strain-softening effect even observed for this last filler [59]. Graft and block copolymer melts were also investigated in the same study and the results clearly showed that rigid-like inclusions yield strain softening behavior, as for hard butadiene particles in poly(acrylonitrile-butadiene-styrene) (ABS) or blocks in poly(styrene-*block*-butadiene-*block*-styrene) (SBS) below its transition temperature, in which case the melt exhibits a two-phase structure. Interestingly, the hardness of the butadiene particles in ABS controls the rheological behavior, from strain hardening with soft particles to strain softening with harder ones. All these studies concluded that the reduction and even elimination of SH could be imparted to the suppression of the large deformation of matrix polymer chains around the particles and the complex flow that may involve shear components between the particles.

Surprisingly, as the dimension of the filler is reduced to a nanoscale, the opposite trend prevails, i.e., adding nanofiller particles can induce SH, as reported for a nanocomposite polymer system made of polypropylene modified with maleic anhydride and montmorillonite clay [60]. A level of clay as low as 4 wt% was sufficient to generate a significant SH response. Higher melt tension and longer drawability of polyethylene–clay nanocomposites were also correlated with the aspect ratio of the nanoclays [61]. The clay particles are usually composed of platelet-like individual layers with a thickness dimension close to 1 nm but with a lateral dimension ranging from several hundred nanometers up to 1 μm, which means a high aspect ratio and a high surface area. Improvement of mechanical properties related to the inclusion of nanoparticles is strongly dependent on the quality of their dispersion through the exfoliation process. Optimum exfoliated clay content, as reported for the poly-ε-caprolactam/montmorillonite clay system, lies in the 1.6 wt% range [62]; above this number the strong interaction between the particles and their bonded polymer molecules may lead to agglomeration of the clay platelets, which is detrimental to the expected enhancement of the end-properties. With well-exfoliated platelets, and in absence of agglomerates at low clay content, nucleation remains homogeneous as in a neat polymer, as deduced from foaming results reported by Nam et al. [63]. This is in agreement with the critical size of an efficient nucleating agent, in the neighborhood of 100 nm, as determined by Ramesh et al. [64]. The foaming experiments performed on nanocomposite systems also illustrated that the platelets self-aligned parallel to the cell walls, with this alignment induced by the strong biaxial extensional flow occurring at the vicinity of the cell surfaces [65].

2.4 Effect of PFA on Rheological Behavior

2.4.1 Reduction of Shear Viscosity

The degree of plasticization achieved is dependent upon many variables, such as types of polymer and PFA, PFA content, temperature, etc. Figure 2.24 displays the viscosity reduction obtained with different foaming agents, and for two types of resin, an amorphous (polystyrene) and a semicrystalline (polypropylene) polymer. The viscosity reduction was calculated from the ratio of the viscosity experienced by the PFA–polymer mixture with respect to that of the neat polymer, both measured at the same constant stress level. While viscosity can be roughly reduced by a factor of two in the case of a semicrystalline polymer, the plasticizing effect of the PFA on the viscosity of polystyrene (PS) is tremendous, and it consequently enables the extrusion process to be run at temperatures very close to the glass transition temperature of the neat PS. In the case of semicrystalline polymers, however, even though the crystallization temperature is lowered by a few degrees due to the presence of the PFA [66,67], the processing window remains practically unchanged as compared to that of the neat polymer.

The plasticization of glassy polymers can be directly translated into the depression of the glass transition temperature [68], which also impacts the viscosity of the molten polymers.

This can be illustrated by the examples shown in Figure 2.25(a) for mixtures of PS and various content of HCFC-142b, where viscosity increases asymptotically as the temperature gets closer to the shifted glass transition temperature, a function of the PFA content (Figure 2.25[b]). In practice, Figure 2.25(a) tells us that PS mixed with 10 wt% of HCFC-142b and processed at 140°C will have the same viscosity as the neat PS at 185°C. The difference in the processing temperatures is approximately the same as the T_g depression displayed in Figure 2.25(b), the small difference being a result of the dilution effect of the PFA acting as a solvent [1].

A fair estimate of the depressed T_g for a diluent weight fraction of w can also be provided by the use of a theoretical relation developed by Chow on a lattice model [69], based on the respective molecular weight of the diluent (M_d) and the polymer repeat unit (M_p):

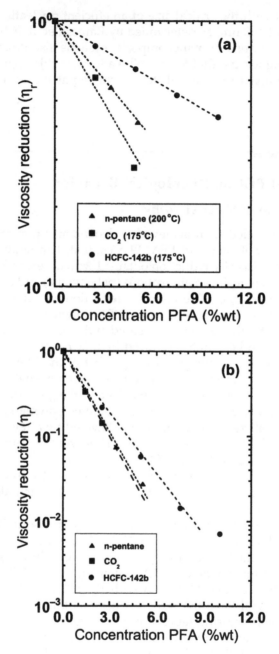

FIGURE 2.24

Viscosity reduction, measured at constant stress, as experienced by (a) PP and (b) PS, with increasing content of various PFAs. ([a] From Gendron, R. and Champagne, M.F., *J. Cell. Plast.*, 40, 131, 2004 [71]. With permission.)

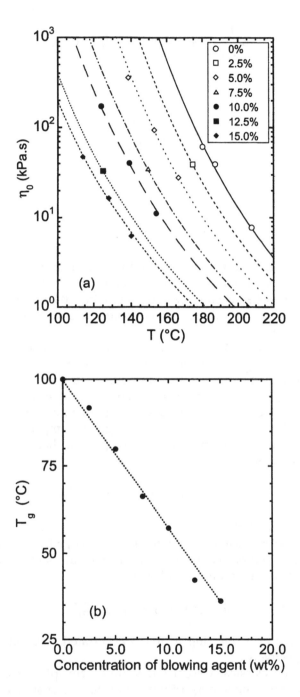

FIGURE 2.25
(a) Zero-shear viscosity values for mixtures of PS and HCFC-142b at different compositions and temperatures, and fitted with Equation 2.7. (b) Glass transition temperatures for the PS/HCFC-142b mixtures as a function of the PFA content, calculated from results displayed in (a).

$$\ln \frac{T_g}{T_{g0}} = \Psi[(1-\theta)\ln(1-\theta) + \theta\ln\theta] \tag{2.19}$$

with

$$\theta = \frac{w/M_d}{z(1-w)/M_p} \tag{2.20}$$

and

$$\Psi = \frac{zR}{M_p \,\Delta C_p} \tag{2.21}$$

where T_{g0} is the glass transition temperature for the pure polymer, T_g is the glass transition temperature of the mixture, ΔC_p is the change in the specific heat of the polymer at its glass transition, z is the lattice coordinate number, and R is the gas constant. Values for these parameters can be found in Chow [69] and Chiou et al. [70].

The rheological signature of mixtures of a polymeric matrix and a physical foaming agent is basically the same as that of the polymer, with the relaxation times shifted to lower values due to the plasticizing effect of the dissolved gas or liquid [1], which is similar to the effect of higher temperatures. This translates into a decrease of the specific relaxation time τ_p through essentially a reduction of the friction factor ξ_0 and a negligible impact on mean-square end-to-end distance per monomer unit a [11]. For a change in temperature, the relaxation times are shifted by a_T:

$$a_T = [\tau_p]_T / [\tau_p]_{T0} = [a^2 \xi_0]_T \, T_0 / [a^2 \xi_0]_{T0} \, T \tag{2.22}$$

In this case, even if the temperature T enters explicitly as a factor, the temperature dependence is mostly due to the change in ξ_0, as in the case of a concentrated solution, where relaxation times are shifted by a_c through dilution and increase of the free volume:

$$a_c = [\tau_p]_c / [\tau_p] \, \rho = [\xi_0]_c / [\xi_0]_\rho \tag{2.23}$$

with indices c and ρ referring to the concentrated solution and the neat polymer, respectively.

Thus applying classical viscoelastic scaling principles, one can perform time-PFA concentration superposition curve shifting analogous to the time-temperature superposition procedure, as displayed in Figure 2.26, with η_0 being the zero-shear viscosity associated with the Newtonian plateau and σ_0 the characteristic shear stress (equal to η_0/τ), this latter being also slightly reduced due to lower level of entanglements in presence of a diluent.

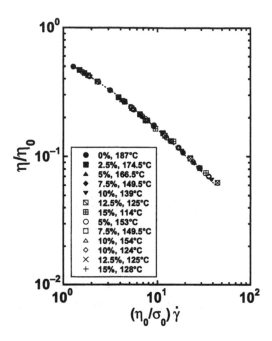

FIGURE 2.26
Master curving for viscosity results obtained with various amounts of HCFC-142b in PS and different temperatures, using appropriate reduced variables.

Results for viscosity modification have been shown in Figure 2.24(a) for various types of PFA dissolved in polypropylene. Figure 2.27(a) now displays viscosity reduction for various polyolefins (POs) combined with the same PFA (HCFC-142b). A different slope is obtained for each polyolefin. Combining Figures 2.24(a) and 2.27(a), it was determined that the viscosity reduction was enhanced as the ratio between the molecular weight of the repeat unit of the resin and the molecular weight of the PFA was increased [71]. In addition it was possible to establish a correlation between the level of plasticization induced by the PFA on different PO and the energy of activation of each type of polyolefin, with the ranking in the energy of activation being as follows: HDPE < LLDPE < PP.

A modification to the approach developed by Chow [69] based on a molecular interpretation of the plasticization for polymer-diluent systems, was also proposed by Gendron and Champagne [71]. A coordination factor for the PFA was introduced, z_{PFA}, and the coordination factor z_P was set equal to 2 for the polyolefin resins. This gave a revised reduced concentration θ, slightly different from Equation 2.20:

$$\theta = \frac{M_P/z_P}{M_{PFA}/z_{PFA}} \frac{w}{(1-w)} \qquad (2.24)$$

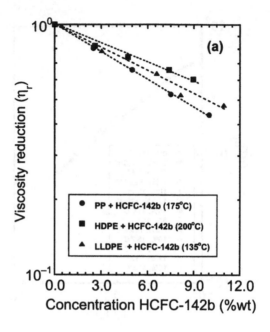

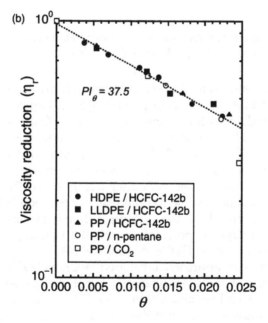

FIGURE 2.27
(a) Viscosity reduction observed for different polyolefins mixed with various concentrations of HCFC-142b, and (b) master curve obtained using a reduced concentration as defined by Equation 2.24. (From Gendron, R. and Champagne, M.F., *J. Cell. Plast.*, 40, 131, 2004 [71]. With permission.)

where the factor z_{PFA} was set to a value of 1 for each foaming agent tested, except the n-pentane (z_{PFA} set equal to 1.4), which also exhibits, to some extent, a linear structure with a repeat unit that can occupy more than one site per molecule.

For cases where a fraction of small branches were randomly distributed along the main backbone, LLDPE for instance, the average monomer-unit molecular weight required by the model was set equal to the same numerical value as that found for the energy of activation. This procedure was based on the observed good agreement between the repeat unit molecular weight (in g/mol) and the energy of activation (in kJ/mol) for some polyolefins.

In Figure 2.27(b) a single master curve is obtained when plotting the reduced viscosity as a function of the reduced molar fraction θ as defined in Equation 2.24, for all the measurements reported in Figures 2.24(a) and 2.27(a). Defining a viscosity reduction index VRI as

$$VRI = -\delta \ln \eta_r / \delta \theta \qquad (2.25)$$

with η_r being the reduced viscosity, one can obtain a single value of 37.5 for all the polyolefins investigated. Extending this correlation to PS was also attempted, as displayed in Figure 2.28. The coordination factor z_p was set equal to a value of 1 for PS [70]. A good agreement was obtained for both sets of results, except for the mixture of PS with HCFC-142b. In this particular case, the polar HCFC molecules may interact with the aromatic rings of the PS, which could be part of the explanation for the enhanced solubility of the PFA in the PS matrix [72] and the enhanced plasticization effect.

2.4.2 Effect of Physical Foaming Agent on Extensional Rheology

As indicated previously, dynamic rheological testing can provide information on the elasticity of the resin, but pure extensional (or elongational) response can only be investigated using an appropriate rheometer. Unfortunately, both approaches preclude the conditions where the polymer is plasticized by the dissolved foaming agent. The study of the impact of plasticization during extensional flow is therefore limited and restricted to modified classical methods.

Investigation of the impact of PFA on extensional response has been performed on polystyrene plasticized by a high molecular weight alcohol, i.e., 2-ethyl hexanol (2EH), which has a boiling point of 196°C. The low vapor pressure associated with this alcohol prevented premature phase separation during the test based on capillary rheometry.

Degree of plasticization of 2EH on polystyrene under shear deformation was first assessed at two different diluent concentrations using an on-line rheometer attached at the end of a twin-screw extrusion line [73]. Tests were performed at 180°C, which is below the boiling point of 2EH. Unfoamed extrudate samples were also collected for the capillary test, and the remaining

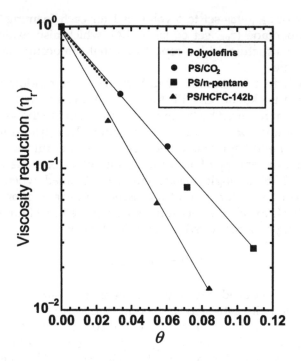

FIGURE 2.28
Results for polyolefins (Figure 2.27[b]) compared to viscosity reduction observed for mixtures of PS and the same PFA, with all the results plotted as a function of the reduced PFA content, Equation 2.24. (From Gendron, R. and Champagne, M.F., *J. Cell. Plast.*, 40, 131, 2004 [71]. With permission.)

alcohol content controlled prior to testing on the capillary rheometer. As displayed in Figure 2.29(a), the two sets of data for the shear viscosity reduction are in close agreement. Moreover, the elongational viscosity as determined from the convergent flow measurement, decreases in presence of a diluent (Figure 2.29[b]), with the same magnitude as that measured under shear deformation. Even though these results provide valuable information relevant to the pressure drop experienced at the entrance of the die due to elongational flow, the conclusions could not be extended to the nonlinear viscoelastic regime, to which the SH effect belongs.

Fortunately, the implementation of a haul-off measurement method at the end of a foam extrusion line yielded valuable results for mixtures of LDPE and hydrocarbons (propane, isobutane, isopentane) as the blowing agents [74,75]. As expected, a decrease of viscosity has been reported with an increase in the PFA content, with the extent of plasticization remaining in magnitudes comparable to those being reported for shear deformations. Although nonlinearity in the viscosity reduction as a function of PFA content was reported, the validity of such measurement is intimately linked to the slow nucleation of hydrocarbon agents (onset of nucleation being a function

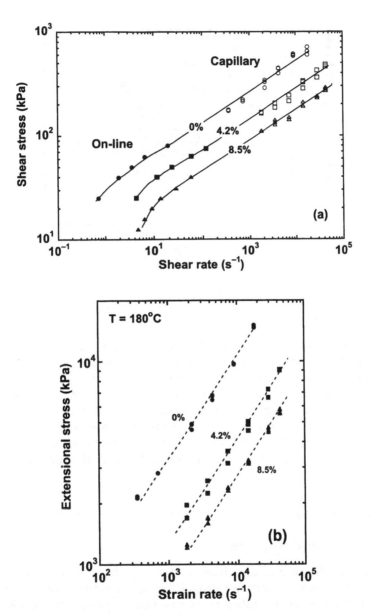

FIGURE 2.29
Comparison of stress reduction of PS in presence of a plasticizer (2EH): (a) Shear stresses measured during extrusion (on-line) and using a capillary rheometer (off-line), and (b) elongational stresses determined from the convergent flow method. All tests performed at 180°C.

of the PFA concentration), since the viscosity measurement should be made in the absence of bubbles in the cross section of the extrudate being deformed.

The plasticizing effect is predominant at smaller strain rates corresponding to a lower draw-down speed (200 rpm — see open symbols in Figure 2.30).

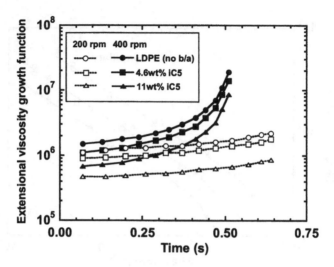

FIGURE 2.30
Influence of isopentane content dissolved in LDPE on elongational behavior, as measured at 130°C at two different roller speeds. (From Ramesh, N.S. and Lee, S.T., *Proceed. Foams '99*, 85, 1999 [74]. With permission.)

However, as the speed is increased (higher rates, draw-down speed set at 400 rpm), the SH effect occurs very rapidly, and the plasticizing effect is essentially present at the earlier times only. For this second set of data, the viscosity responses collapse into one single curve at high times or strain, which may indicate that the foaming agent and its plasticizing effect have less impact on the SH mechanism. It can then be suspected that the long relaxation times associated with the SH behavior are less affected by the PFA molecules than the linear viscoelastic regime driven by smaller relaxation times.

2.5 Failure and Rupture

Especially in the case of low-density foaming, the cells must sustain very high deformations without rupturing. Interest in conditions of rupture might also be driven by the manufacture of open-celled structures, as this feature controls the cushioning properties of the end-product.

Several indicators can be proposed, based on the ability of the melt to sustain either high loading or large deformation, such as the magnitude of SH or the melt-strength value. In fact, the melt strength as measured with a haul-off apparatus is directly related to the limiting conditions under which rupture is observed under uniaxial deformation. It has been previously reported that for polyethylenes, irrespective of long chain branching and molecular weight, the melt strength is inversely proportional to the maximum stretch ratio, with this dependence being a unique function regardless

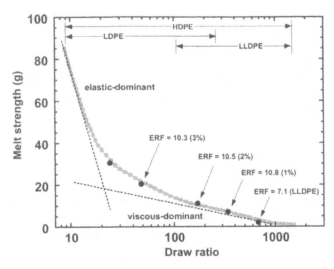

FIGURE 2.31
Master curve for the melt strength as a function of draw ratio for polyethylene resins. (Adapted from Romanini, D., *Polym.-Plast. Technol. Eng.*, 19, 201, 1982 [76]), with examples for expandable blends of standard LLDPE and crosslinked LLDPE (percentage indicated in parentheses, along with the expansion ratio of the foams [ERF]). (Data from Yamaguchi, M. and Suzuki, K.I., *J. Polym. Sci.: Part B: Polym. Phys.*, 39, 2159, 2001. [77].)

of the type of polyethylene [76]. However, it is well known that LDPE and HDPE will not perform the same way during extrusion foaming, and SH behavior as observed for branched polymers remains a critical parameter for foam stability.

Nevertheless, the relationship between melt strength and maximum stretch ratio is schematically represented in Figure 2.31. The curve was sketched around the data points excerpted from Yamaguchi and Suzuki [77]. In that study, small portions of crosslinked LLDPE were compounded with LLDPE to enhance both melt strength and foamability. The low value of the expansion ratio of foam (ERF) of LLDPE was explained by cell wall rupture and coalescence of the cells leading to a bimodal cell distribution. Increasing the melt strength through blending with crosslinked LLDPE helped to stabilize the cell structure prior to any cell breaks. However, expansion was gradually reduced as the gel content was increased.

Since the onset of the rupture mechanism could be related to any inhomogeneity in the deformation field, concentrating load in the thinnest zones, it is worthwhile to mention a study performed on the extensional performance applied to the blow molding process, for which homogeneity of deformation controls the uniformity of the parison wall thickness [78]. Comparison was made between two HDPEs that exhibited process performances opposite to their SH magnitude, as observed under uniaxial deformation. The authors found that the relevant indicator should then be based on the absolute maximum value of the stress at yield, as measured under biaxial deformation.

In this section, many terms will be used to define the damage occurring to the polymer being stretched. The term *failure* is usually associated with a decrease in the material resistance to extension, which leads to yielding or *necking*, this latter term being defined as a localized reduction of the filament's cross section. In that case, the mode of rupture will be ductile. *Cohesive fracture* is associated with a brittle rupture that does not involve infinite reduction of the filament's cross section.

2.5.1 Homogeneous Deformation: The Considère Criterion

The Considère criterion is used in solid mechanics to predict the onset of necking of a material submitted to tensile deformation. Applied to polymer melt, the development of this criterion indicates that the uniaxial elongation of a viscoelastic filament will remain homogeneous until a maximum is obtained in the force curve function of the deformation strain [79]. Written in terms of the transient Trouton ratio Tr^+, the ratio of the transient elongation viscosity η^+_E to the zero-shear viscosity η_0, the condition for homogeneous deformation can be stated as follows:

$$d\,Tr^+/d\varepsilon - Tr^+ \geq 0, \quad \text{or} \quad d\ln Tr^+/d\varepsilon \geq 1 \qquad (2.26)$$

The conditions under which the above criterion may be fulfilled are referred to in several publications as *failure stress* and *failure strain*, which is in conflict with the previous definition of failure that is reserved for the onset of necking. In the present text, *Considère stress* and *Considère strain* will be used instead for the conditions that satisfy Equation 2.26. Fortunately the term *rupture* has been always reserved for true severe damage on the polymer sample (i.e., breakage).

Figure 2.32 illustrates the application of the Considère criterion to different polymers, one linear PP known to be unsuitable for foam application, one branch-modified PP that exhibits at all strain rates the desired SH feature (as shown previously in Figure 2.7), and a blend of the two (85 wt% of linear with 15 wt% of branched PP). Plotting Equation 2.26 as a function of the strain, we find that the three polymers surprisingly share the same left portion of the curve (small strains), where the Considère criterion is rapidly met at a Hencky strain lower than 0.5. This portion of the response belongs to the linear viscoelastic behavior. However, the curves for the branched PP and the blend remain flat and very close to the condition fulfillment of the criterion ($d\ln Tr^+/d\varepsilon = 1$), due to the SH behavior associated with the presence of a branched architecture. For these cases, we can expect that the deformation will remain homogeneous over the entire strain range. However, for all samples, no rupture was observed during the test that was conducted up to a Hencky strain value of 3. On the contrary, the curve for the linear PP kept decreasing, and true necking of the cylinder-shaped polymer sample, as determined from the location of the maximum in the stress response, was obtained at a strain of 2 (and at $d\ln Tr^+/d\varepsilon \approx 0$).

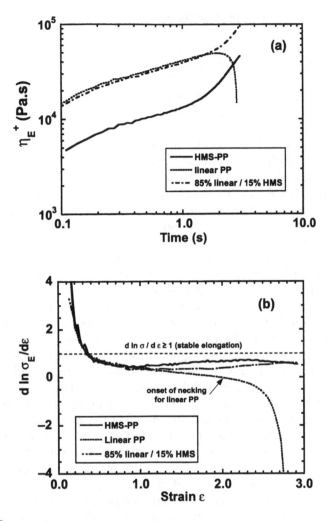

FIGURE 2.32
(a) Transient elongation viscosity measured at a strain rate of 1.0 s^{-1} and 180°C for a linear PP, a branched PP, and a blend of the two PPs. (b) The same results, but after conversion based on Equation 2.26, as a function of strain.

2.5.2 From Liquid to Glass-Like Behavior

The most extensive and comprehensive work on the rupture of polymer melts in extension, which gathered all the published results into a unified theory, was published in 1997 [80]. As noted at the beginning of this paper, "Rupture is one of the least investigated and least understood features of the rheological behaviour of polymeric liquids," although many processes such as blow molding, film blowing, fiber spinning, and extrusion foaming are dominated by extensional flows. Since then, the rupture behavior of melts has increased slightly in popularity, though only a limited amount of work is now being generated.

Pioneering works of Vinogradov and coworkers [81] based on polyisoprenes and polybutadienes of very narrow MWD already showed that polymers undergo a transition from a fluid state, having a large deformation before rupture, to a rubbery state involving limited elastic deformation. Brittleness and reduced deformation characterizes the last, glassy-state regime.

In the same period, the works of Ide and White [82] on the failure of polymer melts under elongational flow were based on commercial resins: (LDPE, HDPE, PP, PS, and poly[methyl methacrylate] [PMMA]). They observed that HDPE and PP exhibit ductile fractures, while cohesive fracture was more generally encountered for LDPE. They have reported that filament stability as well as elongation-to-break depend upon the Weissenberg number, $Wi = \tau_m \dot{\varepsilon}$, with τ_m being the maximum relaxation time. In addition, break conditions for LDPE were found to be highly dependent on the rate of deformation, which motivated the introduction of a deformation rate dependency parameter for the relaxation times.

The work of Pearson and Connelly [83] confirmed the validity of the Weissenberg number to be an indicator of the limit of uniform stretching, as determined from the application of the Considère criterion. In addition, the tensile stress-at-break exhibited a good correlation with the normalized time-to-break (t_B/τ_M). Maximum stresses were obtained with a small t_B/τ_M value, then typically with polymers exhibiting very long relaxation times.

The paper published in 1997 by Malkin and Petrie on rupture based on uniaxial stretching [80] followed and detailed the proposal of Vinogradov based on the transition from a viscous fluid state at a low strain rate up to a glassy state at a higher rate, with an intermediate rubbery, elastic state. Scaling of the rate could take into account molecular weight and temperature, through the maximum relaxation time τ_M, as in classical rheological practice. The Weissenberg number is thus proposed as a parameter to locate transition from one regime to the other. Four regimes were identified as a function of magnitude of rate (or Wi), and they are summarized in Table 2.2.

As indicated in Table 2.2, the departure from the flow regime into the transition zone occurs when $\varepsilon_0 = 0.5$, with ε_0 being the elastic (or recoverable) strain, and not the total strain that also includes irreversible deformations. This corresponds to a critical stress $\sigma_0 = 0.18$ MPa, as reported for several polymers (monodisperse and linear). With the elastic component increasing for higher Wi, the stress at break σ^* becomes a function of the elastic strain at break ε^*_e [80]:

$$\sigma^* = \sigma_0 + 1.2\ (\varepsilon^*_e - \varepsilon_0)\ [\text{MPa}] \qquad (2.27)$$

This study was limited to linear polymers, but it can be anticipated that branched polymers would follow a similar path, with an obvious shift to lower strain rates compared to their linear counterparts, due to their inherent

TABLE 2.2

Different Types of Rheological Behavior Experienced Under Uniaxial Stretching, as Proposed by Malkin and Petrie [80]

Regime	Flow (I)	Transition (II)	Rubbery (III)	Glass-Like (IV)
Rates	Very low rates ($Wi < 0.5$)	Low rates	Intermediate to high rates	Very high rates
Type of deformation	Very large irreversible deformation	—	Limited reversible deformation	Reduced deformation
Type of flow	Viscous flow	Viscous flow and elastic deformation	No flow: fully elastic	No flow and no elasticity
Failure/rupture mode	Yield (necking) but no rupture	Break before steady flow	Failure and rupture are simultaneous	Brittle rupture (cohesive failure)
Limiting strain	Virtually infinite strain[a]	Small and limited strain	Increase of limiting strain	Decrease of limiting strain

[a] In this case, the rupture will be controlled by surface tension. In fact, for a Newtonian fluid, it has been demonstrated that the filament should not fail in finite time [84].

elastic-like behavior. It has been reported for entangled branched polymers that for rapid stretching that leads to the rubbery region, in which the external work applied is entirely stored elastically, the strain to failure is no longer a function of the strain rate [79]. This critical Hencky strain-to-rupture then becomes equal to the failure strain as defined by the Considère criterion, described previously. The onset of instability is also shifted to higher strains with the increase of LCBs.

Interestingly, the experimental results of Barroso and Maia [85] on the strains at failure and rupture as observed for a given LDPE validate the previous theoretical findings of McKinley and Hassager [79]. Moreover, they showed for polyisobutylene (PIB) and PP convergence of both strain-at-failure and strain-at-rupture with increasing strain rate.

2.5.3 Effects of Temperature, Molecular Weight, and Rate of Deformation on Failure of PS Samples

To complement and illustrate these theories, we report here the results of a study conducted on the failure mechanism as experienced by ten different foamable polydispersed linear PS resins submitted to uniaxial elongational deformation using the Rheometric Elongational Rheometer™ (RER). For this study, the parameters investigated were the molecular weight (zero-shear viscosity η_0 determined at 180°C scaling between 9.8 and 172.0 kPa × s), the temperature (from 150 to 185°C), and the rate of deformation (from 0.1 to 1.0 s^{-1}). Attention has been paid to the onset of possible failure based on

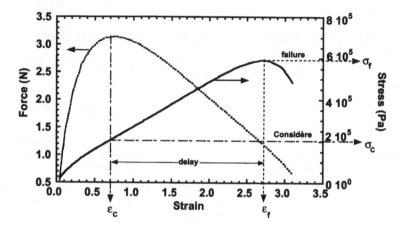

FIGURE 2.33

Determination of the critical stresses and strains for a PS resin tested under uniaxial deformation (170°C, 1.0 s^{-1}), according to the criteria of Considère (maximum in force) and failure (maximum in stress).

the Considère criterion, and the observed "true" failure that corresponds to the necking of the filament, as determined from the abnormal decrease of the elongational stress. For these two events, referred to in the text as "Considère" and "failure," both critical stresses (σ_c and σ_f) and strains (ε_c and ε_f) have been determined from the transient elongational response as shown in Figure 2.33.

The results are summarized in Figures 2.34 and 2.35. The stresses based on the Considère criterion follow a single dependency with the strain as shown in Figure 2.34, irrespective of the temperature or molecular weight. Stress increases steadily as a function of strain, with a break in the linear dependency at $\sigma_c \approx 0.2$ MPa, which is very close to the critical stress defined previously for the transition from a flow regime to the transition one. In addition, we can see that the intercept of the slope at the lower strains corresponds to the stress that marks the departure from the Newtonian regime. Results based on the failure criterion exhibit much more scatter, and may require a closer examination (as detailed below). Nevertheless, it can be expected that the stresses and strains as determined from the two criteria will collapse into a single dependency at higher stresses, typical of the rubbery region (Malkin-Petrie [M-P] regime III). In summary, all the results belong to the viscoelastic regime, but only those lying at stresses above 0.18 MPa, i.e., those that have a significant stored elastic component, would be part of the M-P regime II (transition zone, Table 2.2). Below that critical stress, viscous flow should dominate over the minor elastic contributions.

Surprisingly, the failure stress is found to be proportional to the stress determined according to the Considère criterion, as shown in Figure 2.35. A one-to-one equality would be observed for stresses that belong to the Newtonian flow regime. Even for the viscoelastic behavior in which the elastic component is moderate, a quasi-complete relaxation would be expected

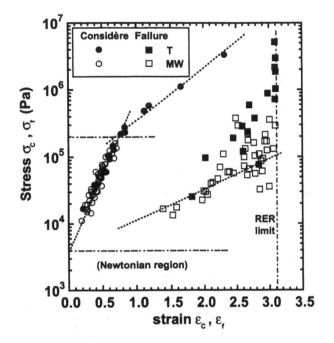

FIGURE 2.34
Summary for the results obtained with ten different PS resins and various testing temperatures. The end of the Newtonian plateau is indicated, as well as the maximum deformation achievable with the elongational rheometer.

during the transient deformation. Examples for this behavior are displayed in Figures 2.36(a) and (b). The maximum in the measured force is rapidly met at short deformation. A plateau in the stress is observed over a wide strain range, as is typical of a steady flow.

The example shown in Figure 2.36(c) also belongs to the viscoelastic regime, but with a significant elastic contribution. The failure of the sample is how-ever observed before any steady flow occurs, and the stress at failure is much greater than the one predicted by the Considère criterion. Surprisingly, if the same sample is tested at a lower temperature, the shapes of the force and stress curves are totally modified, as illustrated in Figure 2.35(d). The "bump" surrounding the maximum force value is shifted to a higher strain (or delayed if considering a timescale). That delay impacts the stress curve. The small transient stresses occurring at small strains (small times) can be associated with viscous dissipation. An increase in the stress at higher strains would be due to the elastic stored components that dominate in this range of deforma-tion. While viscous terms are time-controlled, elastic components are strain-controlled. It should also be noted from this last figure that the sample did not exhibit any failure during its stretching, which was unfortunately limited to a Hencky strain of 3.1, a limitation imposed by the rheometer.

Similar behaviors have been reported in the case of transient stresses for LLDPEs of various molecular weights and polydispersities during uniaxial

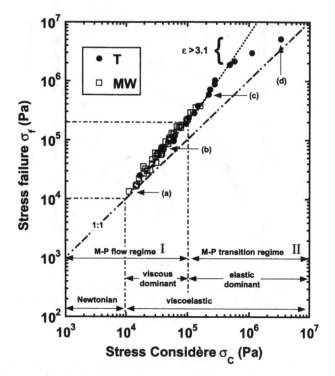

FIGURE 2.35
Critical stresses, failure vs. Considère, for all the PS resins under investigation, with indication of the flow regimes; (a) to (d) indicate samples that have their elongational responses displayed in Figure 2.36.

stretching, using the same elongational rheometer [33]. For low to moderate polydispersity, steady flows could be achieved; these are associated with large strains at failure. Increasing the higher molecular weight content, and thus producing wider MWDs, induced premature failure and rupture at lower strains, with the impact of the polydispersity M_z/M_n being especially noticeable with the increase of the strain rates. This study suggests that the elastic contribution is magnified through the combination of long relaxation times (high molecular weight components that exhibit very long relaxation times are best reflected through M_z) and fast strain rates.

The same study also reported that samples that had been molded over a longer period of time (40 vs. 16 minutes) exhibited higher values of strain at break. This can be associated with the shear history experienced at the compounding and granulation stages, and relaxation (soaked time) on the rebuilding of the original level of entanglement, as reported previously for branched polymers [32].

The scattered results displayed in Figure 2.34 for the failure condition are reproduced in Figure 2.37. Results obtained for several strain rates for each PS and/or temperature are grouped and identified, and a clear common trend emerges from these groups. As the strain rate is increased, stress

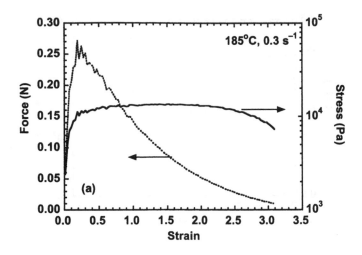

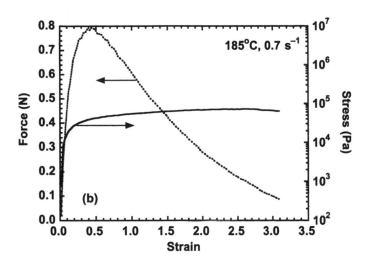

FIGURE 2.36
Different rheological behaviors obtained under uniaxial deformation of PS melts — (a) and (b) illustrate steady-state flows (flow regime, M-P regime I), while (c) and (d) are typical of simultaneous viscous flow and elastic deformation (transition regime, M-P regime II).

increases as well, until a maximum achievable stretch is reached. Further augmentation of the rate of deformation may prohibit adequate relaxation and significant elastic behavior will prevail, restricting the deformation to smaller strains but higher stresses (schematically represented in the upper left corner of Figure 2.37). Although this behavior, which has a right-arrow shape in Figure 2.37, is the same for any set of data, molecular weight and temperature appear to exert a control on the location of this common trend. The appropriate combination would allow a maximum deformation at the

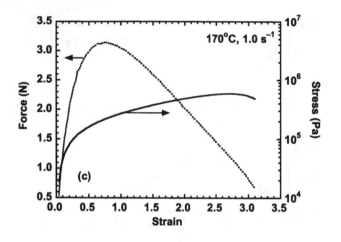

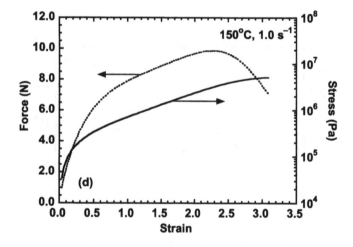

FIGURE 2.36 (CONTINUED)

highest stress, taking into account the rate of deformation dictated essentially by the nucleation and growth conditions (presence of nucleating agents, choice for the foaming agent, etc.).

This approach ought to be extended to semicrystalline polyolefin resins, with the strong expectation that the SH behavior related to elasticity induced by long relaxation times would show a distinct behavioral feature. Preliminary results are shown in Figure 2.38. Linear resins lie close to the one-to-one relationship associated with a near-Newtonian, viscous flow. Resins that exhibit SH (Figure 2.14) deviate considerably from that trend.

In summary, several criteria should be met in order to produce low-density foam with adequate closed-cell morphology. First, the early cell growth following the nucleation stage is only possible with a low-viscosity resin.

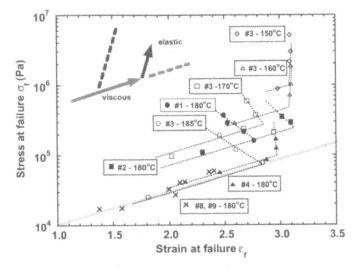

FIGURE 2.37
Critical stresses and strains as defined by the failure criterion, and grouped according to resin sample and temperature. Sample numbers rank the resins from high (#1) to low (#9) viscosity. Each data group consists of several strain rates (increasing with stresses). The schematic in the upper left corner splits the stress/strain result into viscous and elastic contributions.

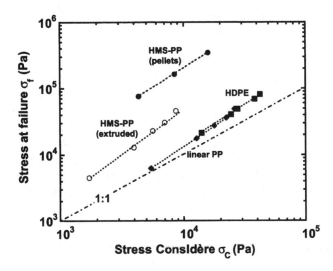

FIGURE 2.38
Critical stresses, failure vs. Considère, for some linear and branched polyolefins.

This criterion obviously excludes resins with a low melt flow rate because of the high resistance to deformation involved.

The second objective is to maintain a homogeneous deformation throughout the cell structure (cell wall thickness), which implies that failure should be delayed to maximum strains. Based on observations made with the

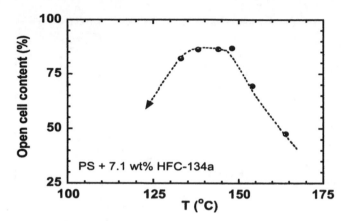

FIGURE 2.39
Open-cell content for extruded foams using PS and HFC-134a, at different extrusion temperatures.

Considère criterion, the conditions for failure are postponed with increasing elastic contribution.

The last objective is to allow maximum expansion while preventing rupture of the cells. Thus subsequent cell growth should be associated with a low Weissenberg number, i.e., for a slow expansion the relaxation time spectrum of the resin could exhibit relatively high values, but if the expansion is "explosive" (high strain rates), high relaxation times are prohibited. The expansion rate could be controlled through an adequate combination of several physical foaming agents.

In addition, the temperature at which the expansion occurs plays a significant role, as illustrated in Figure 2.37. Lowering the temperature increases the strength of the melt (stress at failure) for a given resin, and can also shift the maximum stretch (strain at failure) to higher values, for a constant strain rate. As an example, Figure 2.39 relates the open cell content to the extrusion temperature, for a mixture of polystyrene with HFC-134a. In this case, lower temperatures would have been required to minimize the rupture of the cells, which was prohibited by the insufficient plasticization induced by the moderate content of PFA (limitations imposed by the solubility of the HFC in PS).

2.7   Conclusions

Obviously, rheology plays a significant role at every stage during the extrusion foaming process, starting with the earlier plasticization of the melt induced by the physical foaming agent, which enables very low extrusion temperatures in the case of amorphous polymers. Shear and extensional flows prevail inside the die, and dictate the pressure drop that should be sufficient to prevent any premature nucleation. The nucleation stage may also be sensitive to the viscoelastic behavior of the polymer. It has been demonstrated that shear impacts the nucleation process; increasing shear rate results in enhanced nucleation cell density [86]. Attempts to relate the viscoelastic behavior of the resin to nucleation are briefly covered in Chapter 5.

But more important is the extensional flow prevailing during cell growth. While the stabilization mechanisms, assumed to be linked to the depletion in foaming agent and rapid viscosity increase, are more straightforward in the case of PS and other amorphous polymers, an adequate foaming processing window for semicrystalline polymers relies entirely on so-called SH behavior, present mainly in branched polymer structures. This behavior can be associated with very long relaxation times that induce during the relatively fast deformation step a pseudo-crosslinked network, with an elastic dominant response at high strain.

Changes in the flow regime, controlled by the maximum relaxation time vs. the deformation rate, evolve from a Newtonian (viscous) flow to a viscoelastic behavior, which impacts the maximum deformation that can be achieved in addition to the stress that the stretched filament or film can support before rupture. Despite their reported high stretchability, purely viscous fluids yield a collapsed cellular structure. An adequate amount of elasticity should be present to support the deformation, while not restricting too much the deformation required for low-density foaming. If the fluid is elastic-dominant, little expansion can be envisaged, and brittle rupture follows (open cells).

So far, SH has been qualitatively identified as the required property for making good foams. More work remains to be done to determine quantitatively the amount of SH required. Moreover, the relaxation times need to be coupled to the deformation rate to determine the achievable level of SH. This suggests that the nucleation/diffusion stage, controlled by the nature of the blowing agent, will definitely impact the foamability of the resin: fast diffusing blowing agents like carbon dioxide will be prone to induce rupture, while slow diffusing ones like hydrocarbons are more suitable for large expansion, which is validated by foam industrial practices. Progress in understanding the extensional behavior and its associated rupture mechanisms is unfortunately compromised by the limited resources (hardware) and technical difficulties experienced during extensional characterization.

Abbreviations

2EH:	2-Ethyl hexanol
ABS:	Poly(acrylonitrile-butadiene-styrene)
BKZ type:	Bernstein-Kearsley-Zapas type

bPP:	Branched polypropylene
DCP:	Dicumyl peroxide
EHDM:	Ethylene-1–hexene–ethylidene norbornene
E-MA-GMA:	Ethylene–methyl acrylate–glycidyl methacrylate
EPDM:	Ethylene–propylene–diene monomer
ERF:	Expansion ratio of foam
GPC:	Gas-phase chromatography
HCFC:	Hydrochlorofluorocarbon
HDPE:	High-density polyethylene
IV:	Intrinsic viscosity
LCB:	Long-chain branching
LDPE:	Low-density polyethylene
LLDPE:	Linear low-density polyethylene
LVE domain:	Linear viscoelastic domain
MFI:	Melt flow index
MI:	Melt index
M-P:	Malkin-Petrie
mPE:	Metallocene-type polyethylenes
MWD:	Molecular weight distribution
PC:	Polycarbonate
PET:	Polyethylene terephthalate
PFA:	Physical foaming agent
PI:	Polydispersity index
PIB:	Polyisobutylene
PMMA:	Poly(methyl methacrylate)
PO:	Polyolefin
PP:	Polypropylene
PS:	Polystyrene
PVC:	Poly(vinyl chloride)
RER:	Rheometric elongational rheometer
SBS:	Poly(styrene-*block*-butadiene-*block*-styrene)
SCB:	Short-chain branching
SH:	Strain hardening
TPE:	Thermoplastic elastomers
UHMW:	Ultra-high molecular weight
UPVC:	Unplasticized poly(vinyl chloride)
WLF equation:	Williams–Landel–Ferry equation

References

1. Gendron, R. and Daigneault, L.E., Rheology of thermoplastic foam extrusion process, in *Foam Extrusion, Principles and Practice*, S.-T. Lee, Ed., Technomic, Lancaster, PA, 2000, chap. 3.
2. Kwag, C., Manke, C.W., and Gulari, E., Rheology of molten polystyrene with dissolved supercritical and near-critical gases, *J. Polym. Sci. Part B: Polym. Phys.*, 37, 2771, 1999.
3. Lee, M., Park, C.B., and Tzoganakis, C., Measurements and modeling of PS/ supercritical CO_2 solution viscosities, *Polym. Eng. Sci.*, 39, 99, 1999.
4. Gendron, R., Daigneault, L.E., and Caron, L.M., Rheological behavior of mixtures of polystyrene with HCFC 142b and HFC 134a, *J. Cell. Plast.*, 35, 221, 1999.
5. Constant, D.R., Crystallization effects on foamability of polyethylenes, *Proc. Foams '99*, Parsippany, NJ, 1999, 111.
6. Sammut, P. and Gendron, R., A comparative study of two extensional rheometers: RER versus RME, *Proc. Polym. Process. Soc. 17th Meeting*, Montréal, 2001.
7. Rohn, C.L., *Analytical Polymer Rheology*, Hanser Publishers, New York, 1995.
8. Dealy, J.M. and Wissbrun, K.F., *Melt Rheology and Its Role in Plastics Processing: Theory and Applications*, Van Nostrand Reinhold, New York, 1990.
9. Cogswell, F.N., *Polymer Melt Rheology: A Guide for Industrial Practice*, Woodhead Publishers, Cambridge, 1994.
10. Dealy, J.M., *Rheometers for Molten Plastics: A Practical Guide to Testing and Properties Measurement*, Van Nostrand Reinhold, New York, 1982.
11. Ferry, J.D., *Viscoelastic Properties of Polymers*, 2nd ed., New York, Wiley, 1970.
12. Steeman, P.A.M., A numerical study of various rheological polydispersity measures, *Rheol. Acta*, 37, 583, 1998.
13. Bernreitner, K., Neißl, W., and Gahleitner, M., Correlation between molecular structure and rheological behaviour of polypropylene, *Polym. Test.*, 11, 89, 1992.
14. Graessley, W.W., The entanglement concept in polymer rheology, *Adv. Polym. Sci.*, 16, 1, 1974.
15. Williams, M.L., Landel, R.F., and Ferry, J.D., The temperature dependence of relaxation mechanisms in amorphous polymers and other glass-forming liquids, *J. Am. Chem. Soc.*, 77, 3701, 1955.
16. Münstedt, H., New universal extensional rheometer for polymer melts. Measurements on a polystyrene sample, *J. Rheol.*, 23, 421, 1979.
17. Meissner, J. and Hostettler, J., A new elongational rheometer for polymer melts and other highly viscoelastic liquids, *Rheol. Acta*, 33, 1, 1994.
18. Schweizer, T., The uniaxial elongational rheometer RME – six years of experience, *Rheol. Acta*, 39, 428, 2000.
19. Sentmanat, M., A novel device for characterizing polymer flows in uniaxial extension, *Proc. ANTEC 2003*, Nashville, TN, 2003, 992.
20. Meissner, J., Development of a uniaxial extensional rheometer for the uniaxial extension of polymer melts, *Trans. Soc. Rheol.*, 16, 405, 1972.
21. Wagner, M.H., Bernnat, A., and Schulze, V., The rheology of the rheotens test, *J. Rheol.*, 42, 917, 1998.
22. Münstedt, H., Dependence of the elongational behavior of polystyrene melts on molecular weight and molecular weight distribution, *J. Rheol.*, 24, 847, 1980.

23. Koyama, K. and Nishioka, A., Comparison of the characteristics of polymer melts under uniaxial, biaxial and planar elongational flows, *Proc. Polym. Process. Soc. 14th Meeting*, Yokohama, Japan, 234, 1998.
24. Cogswell, F.N., Converging flow and stretching flow: a compilation, *J. Non-Newt. Fluid Mech.*, 4, 23, 1978.
25. Ahmed, R., Liang, R.F., and Mackley, M.R., The experimental observation and numerical prediction of planar entry flow and die swell for molten polyethylenes, *J. Non-Newt. Fluid Mech.*, 59, 129, 1995.
26. McLeish, T.C.B. and Larson, R.G., Molecular constitutive equations for a class of branched polymers: The pom-pom polymer, *J. Rheol.*, 42, 81, 1998.
27. Doi, M. and Edwards, S.F., *The Theory of Polymer Dynamics*, Oxford University Press, Oxford, 1986.
28. Minegisshi, A. et al., Uniaxial elongational viscosity of PS/a small amount of UHMW-PS blends, *Rheol. Acta*, 40, 329, 2001.
29. Münstedt, H., Viscoelasticity of polystyrene melts in tensile creep experiments, *Rheol. Acta*, 14, 1077, 1975.
30. Messé, L., Pézolet, M., and Prud'homme, R.E., Molecular relaxation study of polystyrene: influence of temperature, draw rate and molecular weight, *Polymer*, 42, 563, 2001.
31. Attala, G. and Romanini, D., Influence of molecular structure on the extensional behavior of polyethylene melts, *Rheol. Acta*, 22, 471, 1983.
32. Yamaguchi, M. and Takahashi, M., Rheological properties of low-density poyethylenes produced by tubular and vessel processes, *Polymer*, 42, 8663, 2001.
33. Schlund, B. and Utracki, L.A., Linear low density polyethylenes and their blends. Part 3. Extensional flow of LLDPEs, *Polym. Eng. Sci.*, 27, 380, 1987.
34. Wood-Adams, P.M. and Dealy, J.M., Using rheological data to determine the branching level in metallocene polyethylenes, *Macromolecules*, 33, 7481, 2000.
35. Wood-Adams, P.M. and Costeux, S., Thermorheological behavior of polyethylene: effects of microstructure and long chain branching, *Macromolecules*, 34, 6281, 2001.
36. Spitael, P., Macosko, C.W., and Sahnoune, A., Extensional rheology of polypropylene and its effect on foaming of thermoplastic elastomers, *Proc. SPE ANTEC 2002*, San Francisco, CA, 2002.
37. Zhang, Z. and Handa, Y.P., An *in situ* study of plasticization of polymers by high-pressure gases, *J. Polym. Sci.: Part B: Polym. Phys.*, 36, 977, 1998.
38. Utracki, L.A., Rheology of poly(vinyl chloride) melts. II. Shear rate-dependent properties, *J. Polym. Sci.: Polym. Phys. Ed.*, 12, 563, 1974.
39. Park, I.K. and Riley, D.W., Elongational flow behaviour of PVC melt, *J. Macromol. Sci.-Phys.*, B20, 277, 1981.
40. Cogswell, F.N., Player, J.M., and Young, R.C., Influence of acrylic processing aids on the extensibility of PVC melts, *Plast. Rub.: Proc.*, 5, 128, 1980.
41. Haworth, B., Chua, L., and Thomas, N.L., Elongational deformation and rupture of rigid PVC compounds for foam extrusion, *Plast. & Rub. Comp. Process. Appl.*, 22, 159, 1994.
42. Champagne, M.F., Gendron, R., and Huneault, M.A., Branched polyethylene terephthalate foaming using HFC-134a: on-line process monitoring, *Proc. SPE ANTEC 2003*, Nashville, TN, 2003, 1870.
43. Japon, S. et al., Foaming of poly(ethylene terephthalate) modified with tetrafunctional epoxy, *Proc. Symp. Porous, Cell. Microcell. Mater.*, MD-Vol. 82, AS-ME, 85, 1998.

44. Graessley, W.W., Glasscock, S.D., and Crawley, R.L., Die swell in molten polymers, *Trans. Soc. Rheol.*, 14, 519, 1970.
45. Xanthos, M. et al., Effects of resin rheology on the extrusion foaming characteristics of PET, *J. Cell. Plast.*, 34, 498, 1998.
46. Gendron, R. and Daigneault, L.E., Continuous extrusion of microcellular polycarbonate, *Polym. Eng. Sci.*, 43, 1361, 2003.
47. Gopakumar, T.G. and Utracki, L.A., Micro-foaming of polycarbonates: a comparison between linear and branched resins. Part 1. Experiments, *Int. Plast. Eng. Tech.*, 4, 1, 2000.
48. Lee, J.W.S., Wang, K.H., and Park, C.B., Challenge to the manufacture of low-density microcellular polycarbonate foams using CO_2, *Proc. Polym.-Supercrit. Fluid Syst. Foams (P-(SF)^2)*, Tokyo, Japan, 212, 2003.
49. Park, C.P. and Malone, B.A., US Patent No. 5,527,573, 1996.
50. Gendron, R. and Vachon, C., Effect of viscosity on low density foaming of poly(ethylene-co-octene) resins, *J. Cell. Plast.*, 39, 71, 2003.
51. Vachon, C. and Gendron, R., Effect of gamma-irradiation on the foaming behavior of ethylene-co-octene polymers, *Radiation Phys. Chem.*, 66, 415, 2003.
52. Schlund, B. and Utracki, L.A., Linear low density polyethylenes and their blends. Part 5. Extensional flow of LLDPE blends, *Polym. Eng. Sci.*, 27, 1523, 1987.
53. Van Calster, M., High melt strength PP foams for the automotive and packaging industries, *Proceedings of the Foamplas '97 Conference*, Mainz, Germany, 1997, 151.
54. Takahashi, T., Takimoto, J.-I., and Koyama, K., Elongational viscosity for miscible and immiscible polymer blends. II. Blends with a small amount of UHMW polymer, *J. Appl. Polym. Sci.*, 72, 961, 1999.
55. Yamaguchi, M., Suzuki, K.-J., and Madea, S., Enhanced strain hardening in elongational viscosity for HDPE/crosslinked HDPE blend. I. Characteristics of crosslinked HDPE, *J. Appl. Polym. Sci.*, 86, 73, 2002.
56. Yamaguchi, M., Rheological properties of linear and crosslinked polymer blends: relation between crosslink density and enhancement of elongational viscosity, *J. Polym. Sci. Part B: Polym. Phys.*, 39, 228, 2001.
57. Le Meins, J.-F., Moldenaers, P., and Mewis, J., Suspensions of monodipserse spheres in polymer melts: particle size effects in extensional flow, *Rheol. Acta*, 42, 184, 2003.
58. Kobayashi, M. et al., Influence of glass beads on the elongational viscosity of polyethylene with anomalous strain rate dependence of the strain-hardening, *Polymer*, 37, 3745, 1996.
59. Takahashi, T., Studies on the effect of macromolecular chain structure on elongational rheology, Ph.D. thesis, Yamagata University, 1996.
60. Okamoto, M. et al., A house of cards structure in polypropylene/clay nanocomposites under elongational flow, *Nano Lett.*, 1, 295, 2001.
61. Wang, K.H. et al., Effect of aspect ratio of clay on melt extensional process of maleated polyethylene/clay nanocomposites, *Polym. Bull.*, 46, 499, 2001.
62. Utracki, L.A., Simha, R., and Garcia-Rejon, A., Pressure-volume-temperature dependence of poly-ε-caprolactam/clay nanocomposites, *Macromolecules*, 36, 2114, 2003.
63. Nam, P.H. et al., Foam processing and cellular structure of polypropylene/clay nanocomposites, *Polym. Eng. Sci.*, 42, 1907, 2002.

64. Ramesh, N.S., Rasmussen, D.H., and Campbell, G.A., The heterogeneous nucleation of microcellular foams assisted by the survival of microvoids in polymers containing low glass transition particles. Part II: Experimental results and discussion, *Polym. Eng. Sci.*, 34, 1698, 1994.
65. Okamoto, M. et al., Biaxial flow-induced alignment of silicate layers in polypropylene/clay nanocomposite foam, *Nano Lett.*, 1, 503, 2001.
66. Dey, S.K., Jacob, C., and Biesenberger, J.A., Effect of physical blowing agents on crystallization temperature of polymer melts, *SPE ANTEC Tech. Papers*, 40, 2197, 1994.
67. Zhang, Z., Nawaby, A.V., and Day, M., CO_2-delayed crystallization of isotactic polypropylene: a kinetic study, *J. Polym. Sci. Part B: Polym. Phys.*, 41, 1518, 2003.
68. Zhang, Z. and Handa, Y.P., An *in situ* study of plasticization of polymers by high-pressure gases, *J. Polym. Sci. Part B: Polym. Phys.*, 36, 977, 1998.
69. Chow, T.S., Molecular interpretation of the glass transition temperature of polymer-diluent systems, *Macromolecules*, 13, 362, 1980.
70. Chiou, J.S., Barlow, J.W., and Paul, D.R., Plasticization of glassy polymers by CO_2, *J. Appl. Polym. Sci.*, 30, 2633, 1985.
71. Gendron, R. and Champagne, M.F., Effect of physical foaming agents on the viscosity of various polyolefin resins, *J. Cell. Plast.*, 40, 131, 2004.
72. Gendron, R. et al., Foam extrusion of polystyrene blown with HFC-134a, *Cell. Polym.*, 21, 315, 2002.
73. Daigneault, L.E. and Gendron, R., Blends of CO_2 and 2-ethyl hexanol as replacement foaming agents for extruded polystyrene, *J. Cell. Plast.*, 37, 262, 2001.
74. Ramesh, N.S. and Lee, S.T., Blowing agent effect on extensional viscosity calculated from fiber spinning method for foam processing, *Proc. Foams '99*, Parsippany, NJ, 1999, 85.
75. Ramesh, N.S., Lee, S.T., and Lee, K., Novel method for measuring the extensional viscosity of PE with blowing agent and its impact on foams, *J. Cell. Plast.*, 39, 281, 2003.
76. Romanini, D., Synthesis technology, molecular structure, and rheological behavior of polyethylene, *Polym.-Plast. Technol. Eng.*, 19, 201, 1982.
77. Yamaguchi, M. and Suzuki, K.I., Rheological properties and foam processability for blends of linear and crosslinked polyethylenes, *J. Polym. Sci. Part B: Polym. Phys.*, 39, 2159, 2001.
78. Tajiri, T., Obata, K., and Kamoshita, R., The effect of elongational properties on thickness uniformity in blow molding, *Proc. Polym. Process. Soc. 14th Meeting*, Yokohama, Japan, 1998, 681.
79. McKinley, G.H. and Hassager, O., The Considère condition and rapid stretching of linear and branched polymer melts, *J. Rheol.*, 43, 1195, 1999.
80. Malkin, A.Y. and Petrie, C.J.S., Some conditions for rupture of polymer liquids in extension, *J. Rheol.*, 41, 1, 1997.
81. Vinogradov, G.V., Viscoelasticity and fracture phenomenon in uniaxial extension of high-molecular linear polymers, *Rheol. Acta*, 14, 942, 1975; Vinogradov, G.V. et al., Flow, high-elastic (recoverable) deformations and rupture of uncured high molecular weight linear polymers in uniaxial extension, *J. Polym. Sci.: Polym. Phys.*, 13, 1721, 1975.

82. Ide, Y. and White, J.L., Investigation of failure during elongational flow of polymer melts, *J. Non-Newt. Fluid. Mech.*, 2, 281, 1977; Ide, Y. and White, J.L., Experimental study of elongational flow and failure of polymer melts, *J. Appl. Polym. Sci.*, 22, 1061, 1978; White, J.L. and Ide, Y., Instabilities and failure in elongational flow and melt spinning of fibers, *J. Appl. Polym. Sci.*, 22, 3057, 1978.
83. Pearson, G.H. and Connelly, R.W., The use of extensional rheometry to establish operating parameters for stretching processes, *J. Appl. Polym. Sci.*, 27, 969, 1982.
84. Hassager, O., Kolte, M.I., and Renardy, M., Failure and nonfailure of fluids filaments in extension, *J. Non-New. Fl. Mech.*, 76, 137, 1998.
85. Barroso, V.C. and Maia, J.M., The influence of molecular structure on the flow and rupture behavior of polymer melt in extension, *Proc. Polym. Process. Soc. 17th Meeting*, Montréal, 2001.
86. Lee, S.-T., Shear effects on thermoplastic foam nucleation, *Polym. Eng. Sci.*, 33, 418, 1993.

3

Polymer Blending Technology in Foam Processing

Michel F. Champagne and Richard Gendron

CONTENTS

3.1 Introduction

The topic of polymer blends has been extensively covered in scientific papers, reviews, and books published over the last twenty years [1–3]. The number of publications is even more considerable if the numerous patents issued on the same subject are considered. Such a durable interest is justified by the large place that polymer blends have secured on the market — approximately 36 wt% of the total polymer consumption, and this number is continuously increasing [4]. Blending polymers is more than ever a widely used industrial practice for a whole range of applications. It is a fast and cost-effective way to customize material properties for specific markets and needs.

Foaming polymer blends present many obvious advantages that will be explored in the present chapter. To date, many commercial blends have been offered, and some have been specially formulated for foaming applications (see Table 1 of Reference 3). But surprisingly, the scientific literature dealing specifically with the use of polymer blends in foaming processes is rather scarce. One of the very first reviews on polymer blends developed for foaming applications was commissioned to the National Research Council of Canada in 1998, as part of the ongoing research program on thermoplastic foaming, FOAMTECH [4]. Since then, part of the information contained in this report has been divulged in monographs devoted to polymer blends [2,3]. The present book chapter does not intend to be as exhaustive as these excellent source books, but will rather be more selective in terms of examples chosen to illustrate the benefits related to processing and applications. Due to the limited information on this topic, the review undertaken here will consist mainly of cases excerpted from the patent literature.

This chapter is divided into two principal parts: (1) fundamentals of blending technology and (2) customizing polymer blends for foam processing and applications. However, it is not the intent of this chapter to provide an in-depth understanding of polymer blending science, and we invite the reader to complement his reading with the abundant and excellent literature devoted specifically to this subject [1–5]. The aim of this chapter is rather to provide basic notions about blending, and to illustrate how its association with foaming technologies can be profitable with respect to improved processability and enhanced performances.

3.2 Fundamentals of Polymer Blending

A polymer blend is defined as a mixture of two or more polymers. However, it is a frequent industrial practice to add a small quantity of a second polymer to modify the processability of the base material. For instance, adding 2 wt% of LDPE to LLDPE is a common practice that enables lower processing pressures and thus increased extruder throughputs. For this reason, the term "polymer blend" is restricted to systems comprising at least 2 wt% of the second polymer. Below that level, the second phase is considered to be an additive [1,2].

The rationale for blending is twofold: it is mostly used to improve material performance while providing manufacturers with cost-effective pathways to achieve a desired set of properties. Blending first originated from the need to improve the impact strength of brittle resins such as polystyrene (PS), poly(vinyl chloride) (PVC), polypropylene (PP), poly(methyl methacrylate) (PMMA), polyethylene terephthalate (PET), polyamide (PA), etc. [5]. But blending has evolved in recent years in a versatile way to control a whole spectrum of material properties: thermal resistance, permeability, biodegradability, strength, toughness, dimensional stability, and so on. However, achieving a full set of targeted performance characteristics is only possible in carefully designed polymer blends. The following sections will briefly present the different concepts involved in the development of polymer blends.

3.2.1 Miscibility

Miscibility is associated with a level of homogeneity down to the molecular scale [5]. This means that no separate domain can be observed since it would be of a dimension comparable to the macromolecular statistical segment. The miscibility of two polymers is thermodynamically characterized by a Gibbs free energy of mixing (ΔG_m) equal to or smaller than zero. Miscible systems should also comply with an additional condition stating that the second derivative of the Gibbs free energy of mixing with respect to composition ϕ should be positive, that is, $\delta^2 \Delta G_m / \delta \phi^2 > 0$ [6]. One of the best-known examples of a commercially successful miscible polymer blend is given by the polystyrene/poly(2,6-dimethyl 1,4-phenylene ether) (PS/PPE) system. This blend is also a good example of a balanced system where each component compensates for an inherent weakness of the other: PS enhances the processability of the intractable PPE, while PPE provides increased thermal resistance to PS. Polymer blend miscibility is unusual, and most common polymers form immiscible systems when they are combined.

Immiscible polymer blends are characterized by positive values of ΔG_m. Separate phases of the two components will coexist at a microscopic or macroscopic scale and will exhibit specific morphologies, as described further in the following sections. It is important to note that most commercial polymer blends are immiscible systems. The phase separation typically exhibited by immiscible systems usually leads to poor blend properties. The challenge here is to modify immiscible blends by inducing some level of interaction between the phase-separated components. This type of modification, typically referred to as *blend compatibilization*, is mostly based on interfacial modification. By carefully controlling the interaction between the phase separated components, it is possible to achieve optimum blend performance. This is illustrated in Figure 3.1, where the morphology and resulting mechanical behavior of 10 wt% elastomeric ethylene-octene (EO) copolymer/PET blends are compared. The control system, prepared from

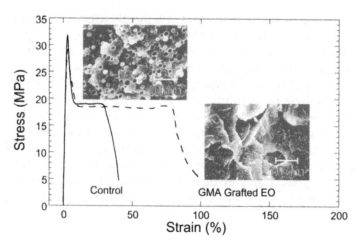

FIGURE 3.1

Morphology and mechanical behavior of 10 wt% EO copolymer/PET blends modified by a glycidyl methacrylate grafted EO copolymer.

unmodified EO copolymer, produced a droplet-type morphology in the 2 to 5 μm range that resulted in a quick mechanical failure of the specimen tested. By contrast, the sample prepared using a glycidyl methacrylate grafted EO copolymer generated a submicron-level droplet morphology. Glycidyl methacrylate moieties grafted on the EO copolymer are known to react with PET end-groups, thus producing a graft EO/PET copolymer that is very efficient in compatibilizing the blend. The resulting compatibilized blend displayed elongation at break values two to three times larger than those reported for the control system. Additional details on the different compatibilization technologies and benefits will be given later in the chapter.

3.2.2 Designing Polymer Blends

3.2.2.1 Targeting Properties

The development of performant polymer blends is certainly a challenging task. Obviously the performance of polymer blends relies on the intrinsic properties of their components, their respective concentrations, and the resultant morphologies and interfacial interactions. Usually, the motivation for blending polymers is driven by the need to compensate for a specific weakness of a given material. For example, one might be interested in reducing the oxygen permeability of a polyolefin for a packaging application. Addition of low O_2 permeability ethylene-vinyl alcohol copolymer (EVOH) and generation of a lamellar morphology will greatly improve the permeability performance of the blend. Depending on specific requirements, different material characteristics will be targeted for the blends' components: low-temperature elastomeric behavior for impact strength improvements, nonflammability for inducing flame-retardancy, high temperature engineering

TABLE 3.1

Selected Examples of Polymer Blend Designed for Specific Requirements

Properties Sought	Generic Requirements	Typical Examples	Ref.
Impact resistance	Droplets of low T_g elastomers	PA/EPDM, PET/EEA	7, 8
Increased mechanical strength	Fibers of high modulus polymer generated *in situ*	PS/PPE/LCP, co-PA/LCP	9, 10
Reduced permeability	Lamellae of low permeability polymers	PE/PA, PE/EVOH, PP/PA	11–14
Moisture resistance	Co-continuous morphology with a water-resistant polymer	PP/PET	15
Biodegradability	Large content of biodegradable polymer; co-continuous morphology works best	LDPE/TPS, TPU/TPS	16, 17
Improved heat resistance	Co-continuous matrix of high T_g polymers; miscible blends with high T_g polymers	PC/PBT, PC/PET, PS/PPE	6, 18, 19

resins for increasing temperature resistance, etc. Selected examples are reported in Table 3.1. Interestingly, some of the examples reported are demonstrating that simultaneous improvement of different properties can be achieved through an adequate selection of components. Blends of semicrystalline material with an amorphous polymer, such as polybutylene terephthalate/polycarbonate (PBT/PC) or polybutylene terephthalate/acrylonitrile-butadiene-styrene copolymer (PBT/ABS), have a better balance of thermal and dimensional stability, chemical resistance, and mechanical properties than their base components. The selection of one resin among several possible candidates finally relies on an adequate equilibrium of their inherent advantages and deficiencies. Since morphology, i.e., optimized size and shape of the dispersed phase, dictates most of the polymer blends' performance, adequate choice of the rheological behavior of the resin should be made, in concert with the proper method of compatibilization, compounding, and processing.

3.2.2.2 Methods of Blending

Mechanical melt blending is undoubtedly the most widely used technology for preparing polymer blends. A precise control of the compounding process variables is critical in order to achieve an adequate dispersion while ensuring that the morphology is well stabilized; otherwise the poor quality of the blend will be reflected through inappropriate mechanical integrity and performance. This is also true for the increasing number of alloys being prepared by reactive processing, where compatibilization is extremely sensitive to the extent of the reaction.

Twin-screw extruders (TSEs) should be the preferred continuous compounding devices. The modular design of TSEs allows manufacturers to customize their processes to meet the specific requirements of each blend they offer. This is especially important in reactive processing, where TSEs provide a lot of freedom in designing the process. For example, very low viscosity chemicals are easily fed and blended in TSEs, different feeding sequences are also allowed, volatile by-products can be efficiently removed, etc. These processing parameters often have a critical impact on blend performance.

Compared to TSEs, single-screw extruders (SSEs) are much less efficient mixing devices. On the other hand, SSEs are more affordable than TSEs. Consequently, they are still very popular, and apparently most of the blends described in the patent literature were prepared on SSEs [4]. The mixing efficiency of SSEs can be easily upgraded through the selection of adequate mixing screws and add-on mixing devices [5], such as the cavity transfer mixer (CTM) [20] or the extensional flow mixer (EFM) [21].

3.2.2.3   Compatibilization

As miscibility is limited to a very small number of systems, it becomes mandatory to ensure compatibility of the phases. Compatibilization aims at modifying the interfacial properties, which is highly critical for the performance of immiscible polymer blends. An efficient compatibilization assists in the development of the required morphology during the compounding stage and helps to preserve this optimized phase structure during the subsequent forming steps. This process is thus a prerequisite for manufacturing polymer blends with stable and reproducible properties. The design of a good compatibilization strategy is the key factor required for achieving high-quality polymer blends. It is usually recognized that an efficient compatibilization method must complete these three tasks:

1. A reduction of interfacial tension between the blends' components to permit the production of a finer dispersion during the compounding step

2. A decrease in coalescence rate to stabilize the morphology during compounding and prevent phase coarsening during the subsequent forming processes

3. An efficient stress transfer between the solid components to provide mechanical strength and integrity to parts manufactured from the polymer blend

The most common compatibilizing technique relies on the use of selected copolymers that are added to the blends. The chemical structure of the copolymer is chosen in such a way that its different segments are selectively miscible with each of the blends' components. The simplest approach is to use a copolymer A–B for compatibilizing a blend of polymer A and polymer B. Of course, different variations on this theme are possible; a copolymer C–D would efficiently compatibilize a blend of polymer A and B, provided that segments C and D are selectively miscible with polymer A and polymer B. But the most efficient and widespread compatibilization technology is based on reactive processing. In this scheme, the copolymer is generated *in situ* during the blending process. The chemical reaction might directly involve the neat components blended together, or selected chemically functionalized analogs added to the polymer system. The transesterification reaction between PC and PET is a good example of the former case, while the addition of maleated PP to the PP/PA blend is a classical example of the latter process. Reactive compatibilization is usually considered more efficient, as it generates the copolymer molecules directly at the blend interface, where they are required for an efficient compatibilization process. Extensive information on polymer blends compatibilization is available in the literature [22,23].

A good example of the different functions expected from an efficient compatibilizer is given in Figure 3.2. There has been a great deal of research in the past few years focused on modifying PET impact resistance by blending technology. The addition of elastomeric particles has been proven efficient in other brittle matrices such as PS and PA. However, the compatibilization of olefinic-based elastomers with PET is complicated by the low reactivity of typical polyester end-groups towards common interfacial modifiers (maleic anhydride and acrylic acid grafted PO). The use of glycidyl methacrylate (GMA) copolymers and terpolymers has been proven an efficient way to circumvent the problem. The GMA moieties efficiently react with PET to generate compatibilizing PET-based copolymers *in situ*. The coarse morphology obtained by blending 10 wt% EO copolymer to PET (Figure 3.2[a]) is reduced to the submicron level upon addition of 2 wt% ethylene, vinyl acetate, and glycidyl methacrylate (EVA-GMA) elastomeric terpolymer (Figure 3.2[b]). Interfacial tension measurements were conducted on PET systems modified using four different GMA-based copolymers. Measurements were made using the well-known breaking thread technique [24,25]. PET fibers were immersed in hot GMA terpolymer solution for various amounts of time, in order to let the species react together and generate a grafted copolymer. The treated fibers were thoroughly rinsed with fresh solvent and embedded in neat EO copolymer films. As shown in Figure 3.2(c), the interfacial tension between PET and EO copolymer was efficiently reduced by the compatibilizing species grafted on the PET fiber surface. The interfacial tension coefficient reached some minimum values when the immersion time was sufficiently long. The different GMA copolymers and terpolymers investigated displayed different levels of interfacial tension reduction, the lightly GMA grafted EO copolymer being the most efficient system. The adhesion enhancement brought by these different compatibilization systems was characterized through simple peel test measurements. Thin films made of the different GMA terpolymers were prepared and

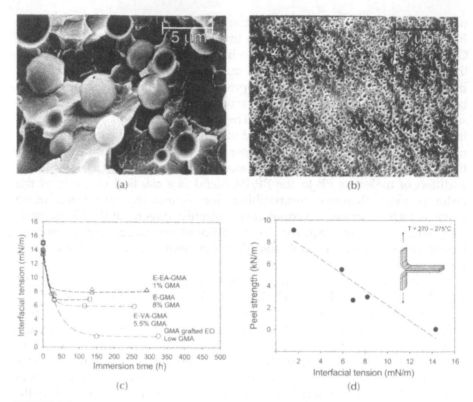

FIGURE 3.2
Compatibilization of PET/EO copolymer blends by GMA copolymers; scanning electron micrographs taken from fracture surfaces of 10 wt% EO copolymer/PET blends are shown in (a) for the control blend and (b) for the system modified with 2.5 wt% EVA-GMA terpolymer. Interfacial tension coefficients measured on compatibilized systems are reported in (c) while the correlation between the peel strength and interfacial tension is given in (d).

embedded between two PET sheets. The assemblies were heated at 270°C to let the copolymers react with PET. In all cases, larger peel strength values were obtained from the modified systems as compared to samples prepared from neat EO copolymer. Interestingly, the peel strength was found to be directly correlated with the minimum interfacial tension reached with these GMA copolymers, as shown in Figure 3.2(d). The species providing the lowest interfacial tension coefficient also generated the best adhesion. GMA-based copolymers in PET/EO copolymer blends then displayed the three basic functions expected from an efficient compatibilizer.

3.2.3 Blend Characteristics

3.2.3.1 Morphology

As underlined previously, morphology is critically related to blend properties. Morphology largely depends on the immiscibility level between the

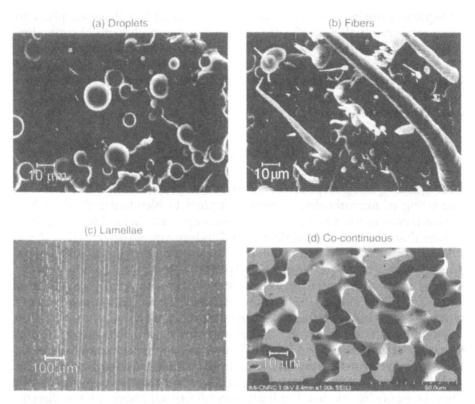

FIGURE 3.3
Scanning electron micrographs taken from the following: (a), (b) Cryogenic fracture surfaces of copolyesteramide/LCP blends [10]; (c) HDPE/PA6 blend microtomed and stained with phosphotungstic acid to reveal PA6 phase structure; and (d) PCL/PEO blends microtomed surface, after PEO water etching to reveal PCL phase structure.

components (as reflected through the interfacial tension) as well as the blend composition and the stress applied on the system. Systems exhibiting large repulsive forces between components such as polyester/polyolefin or polyamide/polyolefin blends display coarse morphological features while blends with mild immiscibility (such as PS/PMMA or PP/EPR) are much easier to disperse.

Examples of typical polymer blend morphology are presented in Figure 3.3. Over the concentration scale, the dispersed phase morphology smoothly evolves from spherical drops to cylinders, fibers, and sheets. Eventually, the coalescence of these larger morphological features will lead to the creation of a continuous phase coexisting with the "matrix" in the middle range of concentration. This co-continuous morphology is usually stable on a fairly wide concentration range. The co-continuity structure is often a desirable feature since it contributes to the synergy of the polymer component properties [3]. Finally, at some larger concentration, the "dispersed" phase will ultimately become the matrix, a phenomenon referred to as *phase inversion*.

Dispersive mixing consists of reducing the size of the dispersed phase to some minimum value mostly dictated by the thermodynamical properties of the system and the stress applied on the blend. One classical mechanism often cited involves the deformation of a droplet under shear or extensional flow fields, until a few conditions are met for the breakup of the deformed drop. Under steady uniform shearing flow, one condition is based on the viscosity ratio λ defined as:

$$\lambda \equiv \eta_d / \eta_m \qquad (3.1)$$

with η_d and η_m being respectively the dispersed phase and matrix viscosities. According to microrheology concepts applied to Newtonian fluids, when the difference in the viscosity of the two components is too large, i.e., for λ greater than about 4, a shear flow would induce simple deformation of the droplet and no subsequent breakup of the resulting filament. As a consequence, dispersion of such high viscosity ratio systems would not be possible using simple shear flow field. This limitation does apparently not apply to extensional flow fields, which would make extensional deformations very attractive for the development of dispersive devices [5].

Polymer melts are visco-elastic fluids and their typical behavior is far from that of idealized Newtonian fluids. They generally display a non-negligible level of elasticity in addition to a mild to severe shear thinning behavior. In addition, flows in typical compounding equipment are far from the simple shear or extension modes. The limitations discussed above are not directly applicable to polymer blending processes and a more complete discussion on this topic can be found elsewhere [26–28]. As a consequence, dispersion of polymer systems with viscosity ratios larger than 4 has been shown to occur in simple shear flows. While dispersion of matched or nearly matched viscosity systems mostly occurs by sheet formation and breakup, blends with high viscosity ratios were rather dispersed through an erosion-based mechanism. Very fine droplets could then be created, even in polymer blends with a viscosity ratio as high as 60 [26]. This mechanism would explain the large and mostly non-dispersed phase (the so-called "gels") usually coexisting with the very fine dispersion often observed in high viscosity ratio polymer blends prepared by extrusion [27].

The addition of a physical foaming agent would be expected to affect the polymer blending process in some way. As mentioned in the introduction to the present chapter, the literature on foaming polymer blends is scarce. A few examples of investigation involving blend compounding in the presence of supercritical carbon dioxide have been reported. CO_2 can be used deliberately as a processing aid to impact the morphology of the blend, as reported for PS–PMMA blends [29]. Dissolution of carbon dioxide in a blend of 75 wt% PS and 25 wt% PMMA, with CO_2 having a strong affinity for PMMA, modified the original dispersed phase size from 1.5 to 0.48 µm upon the addition of the supercritical fluid (SCF). A similar behavior has also been reported for the PE/PS blend [30]. Such morphology modifications through the addition of SCF have been shown to modify the expansion stage of the blend foaming process. Another example reports the foaming of a reactive blend of PP and a copolymer of ethylene and maleic anhydride or ethylene-methyl acrylate (EMA), where the EMA phase is dynamically crosslinked [31]. The addition of CO_2 was shown to have no impact on the crosslinking reaction itself, but it delayed the phase inversion to a higher crosslinking level. In addition, it was shown that the blend morphology played a critical role during the foaming process. First, it should be underlined that the foam cellular structure and the blend morphology belong to two different scales. While the cell size is in the range of 100 to 200 µm, the blend morphology is much finer, in the 1 to 2 µm range, as shown in Figure 3.4. The observed cell structure was typically controlled by the polymer forming the matrix. Since EMA was much easier to foam than PP, lower density foams were obtained when EMA was the matrix phase.

3.2.3.2 *Rheology*

Unlike simple liquid structures that exhibit either Newtonian or non-Newtonian behavior well described by classical rheological theories, the viscosity of polymer blends is concentration-dependent. The viscosity may in many cases follow intricate deviations from a log-additivity rule, $\ln \eta = \Sigma \, \phi_i \ln \eta_i$, and may be either positive, negative, or both. Since immiscible blends, due to their inherent two-phase flow, behave differently from their miscible counterparts, rheological characterization can be used as an adequate tool to assess the miscibility of the components. A complete discussion of the rheological behavior of polymer blends can be found in Utracki and Kamal [32].

The microstructure of the blends is very strain sensitive. Thus the response obtained from dynamic characterization performed under small strain amplitude will be different from that observed in capillary viscosimeter testing, i.e., under large strain steady-state flow conditions. As stated in Reference 5, "in polymer blends the material morphology and the flow behavior depend on the deformation field, thus under different flow conditions, different materials are being tested." While capillary testing is more appropriate to mimic the flow behavior prevailing in the process (flow in a die or during the molding-filling state), dynamic tests, due to their noninvasive nature, will provide reliable information on material characteristics.

The viscosity, elasticity, and yield stress of polymer blends are all increased through compatibilization, since this latter greatly impacts dispersion, interface rigidity, and interactions between the two phases and the dispersed droplets themselves. Compatibilization is also highly beneficial to the extensional rheological behavior of immiscible blends, especially when preserving the strain hardening feature that is mandatory to foam applications [5].

As reported in Chapter 2, the presence of a physical foaming agent (PFA) acts as a plasticizer, with the magnitude of the plasticization effect being a function of both the polymer and the PFA. For a polymer blend, the overall

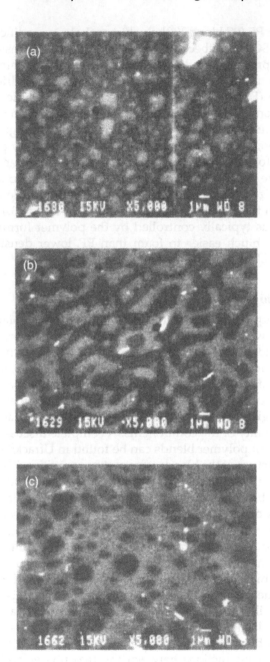

FIGURE 3.4
Morphologies observed in the cell walls of PP/EMA foams blown with carbon dioxide, for various EMA contents: (a) 30 wt% (PP matrix), (b) 50 wt% (PP matrix), (c) 70 wt% (EMA matrix). The EMA phase was stained with phosphotungstic acid and appears as bright zones on the pictures. (From Pesneau, I. et al., *J. Cell. Plast.*, 38, 421, 2002 [31]. With permission.)

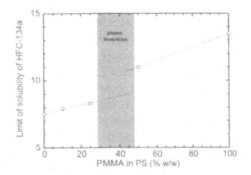

FIGURE 3.5
Increase of the solubility limit of HFC-134a with the PMMA fraction in PS/PMMA blends (Adapted from Smith, P.J. et al. [33]). Discontinuity in the trend could originate from the anticipated phase inversion.

rheological response will be complicated by the presence of the PFA molecules that could induce very different viscosity decreases for the two polymers, as in the case of a PE/PS blend. Moreover, each polymer may not have the same solubility limit with respect to the swollen gas, under given pressure and temperature conditions. This is illustrated with the immiscible PS/PMMA blend modified by HFC-134a addition. This PFA exhibits a solubility limit of 8 wt% in PS and 18 wt% in PMMA (data taken at 150°C, 5 MPa) [33]. Increasing the fraction of PMMA in the blend shifts to higher values the amount of HFC-134a that can be dissolved, as illustrated in Figure 3.5. In addition, the nature of the matrix phase, either PS or PMMA, impacts the overall solubility, as illustrated by the near-parallel lines of different amplitude on each side of the expected phase inversion concentration. One would then expect the magnitude of the viscosity depression experienced in the PMMA phase to be much more severe than the plasticization achieved in the PS phase.

3.3 Customizing Polymer Blends for Foam Processing and Applications

The majority of low-density foams are manufactured from PE, PP, PS, and PVC, all commodity resins. Foaming is performed with either a chemical blowing agent (CBA) or a PFA, such as hydrocarbons, atmospheric gases, or hydrochlorofluorocarbons. Combinations of CBA and PFA are also frequently reported, with the role of the CBA mainly restrained in these cases to nucleating purposes. It is also a common industrial practice to foam styrenic, olefinic, vinylic, and EO blends, as will be highlighted in this section. For example, PVC was blended, and subsequently foamed, with crosslinked acrylonitrile-butadiene rubber (NBR) as early as 1947, in the infancy days of thermoplastic foaming (PS foam was introduced in 1931 and PE foam in 1941). This blend was used for the production of articles such as buoyancy life vests and in applications that require materials having shock-absorbing as well as insulating properties [34]. Foaming is not restricted to commodity resins, and low-density engineering foams can also be produced with specific targeted properties such as higher rigidity, mechanical perfor-mance, and thermal stability [4]. However, the main reason for foaming such expensive engineering and specialty resin blends has been reduction in material cost, through a typical 30% weight reduction of the molded part.

Obviously, one can benefit from the enhanced properties of the blended material, as well as some of the other advantages detailed in Section 3.2. For example, PS is glassy at ambient temperature and thus brittle. PS as well as other amorphous polymers that have a high glass transition temperature may benefit highly from blending if impact strength and ductility improve-ments are being sought [4]. Specific examples will be provided in a subse-quent section, focusing on the physical and mechanical property enhancements achieved in foamed blends.

It should be highlighted that miscible blends are more suited to foaming processes. For example, PS is miscible with PPE, which makes the properties of the foams based on this combination highly dependent on the engineering resin content, with obvious improvement in mechanical strength and tough-ness. The glass transition temperature of the blend increases with the fraction of PPE, from 100°C (PS) to 220°C (PPE). Extruded foams made from PPE and CFC-12 had a density of 96 kg/m^3, and a heat deflection temperature (HDT) of 192°C [35]. The same patent also claimed that blending PPE (20 to 90 wt%) to either PS or high-impact polystyrene (HIPS) increases compres-sive strength for low-density foams (16 to 40 kg/m^3). For example, blending 30 wt% of PPE with 70 wt% of PS, and extrusion foaming with either CFC-11 or CFC-12, yielded foams of densities in the range of 30 to 125 kg/m^3.

Well-compatibilized blends are also particularly attractive (e.g., PS/PE). The absence of compatibilization could yield poor mechanical properties that will be reflected in the foam quality. In cases where compatibilization cannot be done, performance of these blends may remain acceptable in given concentration ranges: when concentration of the dispersed phase is less than 10 wt%, and at the concentration close to the phase inversion, i.e., when the morphology is co-continuous. Immiscible blends should also be character-ized by a high degree of dispersion, as for the blends of different PEs [4].

Blending can provide substantial benefits related specifically to the foam-ing process. Numerous blends were formulated especially to enhance their foamability. Improving processability can be addressed through different issues: (1) increasing the strain hardening effect; (2) providing a better control of the nucleation and growth mechanisms; (3) improving the control of closed/open cell content; (4) modifying the gas solubility and diffusion, which could also impact the dimensional stability. These opportunities will be covered first in the following sections.

3.3.1 Strain Hardening and Foamability

The lack of strain hardening shown by most semicrystalline polymers severely limits their use in foaming processes, the frequently cited examples being the standard grades of high-density polyethylene (HDPE), linear low-density polyethylene (LLDPE), and especially PP. In fact, the processability of PO resins relies on molecular structural characteristics such as molecular weight (MW), molecular weight distribution (MWD) and, more relevant to the foaming process, the degree and type of branching. These characteristics also affect their properties. In fact, the molecular structure controls the crystallinity, which in turn impacts the density and modulus. Linear HDPE ($\sim$ 940 to 970 kg/m^3) exhibits high crystallinity, this latter being shifted to lower levels with the introduction of long chain branching as in LDPE ($\sim$ 915 to 930 kg/m^3).

A solution to alleviate the lack of strain hardening inherent in many linear POs consists in blending the linear polymer with a resin that already exhibits strain hardening. For instance, compared to their homopolymer counterparts, blending LDPE with HDPE offers a simple way to finely control the magnitude of strain hardening, this latter behavior being typical of branched resins (LDPE for example). Also, the mechanical properties can be preserved with a high amount of the linear, high-crystallinity PE (HDPE) in the blend. Unfortunately, polyethylene resins are immiscible with almost any polymers, and immiscibility even prevails between molten PEs having different chain structures. Co-crystallization for these systems is also rare. As the molten polymers undergo crystallization, the first spherulites to appear, associated with the PE that has the higher melting temperature, will subsequently be encapsulated by those of the second PE. Fortunately, since phase coarsening kinetics is controlled by the reduction of the overall surface energy, the low interfacial tension existing between the different PEs will make this process very slow [4]. Thus, good performance can still be obtained from PE blends exhibiting a high degree of dispersion. Otherwise, immiscible blends like HDPE with LDPE may need compatibilization that can be performed by adding adequate copolymers, such as ethylene–propylene elastomeric copolymer (EPR) or LLDPE. In addition to enhanced foamability, other advantages reported for such PE blending are abrasion resistance, stiffness, transparency, stress crack resistance, and so on [4].

Polypropylene (common isotactic PP, with a melting point close to 165°C) is not particularly suited for foaming. The poor foamability of PP is essentially related to the absence of strain hardening in its extensional rheological behavior (see Figure 2.32a). However, foaming PP gained in popularity with the introduction in the early 1990s of a branched, high melt strength PP which exhibited many similarities with LDPE in terms of extensional behavior (i.e., the presence of strain hardening). This branched PP has a similar melting point (160 to 166°C) and is usually blended with its linear counterpart [36]. Because of the frequent mismatch of viscosities, homogeneization of these mixtures may be particularly problematic and needs to be addressed correctly through proper mixing.

The same linear/branched blending strategy was also claimed to be suitable for the foaming process of polycarbonate, for branched PC content ranging from 60 to 95 wt% [37]. The advantages are said to be related to the low viscosity experienced at a high shear rate, and the enhanced melt strength properties inferred from the low-frequency storage modulus.

Improvement of strain hardening for PE blended with either polyisobutene (PIB), atactic PP (aPP), or polybutylene (PB) has been reported as well [1]. In general, blending PE with elastomers (chlorosulfonated polyethylene [CSPE], PB, EPR, elastomeric terpolymer from ethylene, propylene, and non-conjugated diene [EPDM], etc.) improves the low temperature impact strength and elongation [4].

Other routes have been proposed to compensate for the lack of strain hardening, through the generation of long chain branching by heterogeneous grafting, for example. Irradiated blends of PP (10 to 50 wt%), semicrystalline ethylene–propylene (EP) copolymer (5 to 20 wt%) and EPR (40 to 80 wt%) were claimed to be useful for foaming applications [38]. The irradiation step was conducted under an accurately controlled active oxygen content, maintained at less than 40 ppm. Such resulting material (30 wt%) was blended with an irradiated PP (70 wt%) and extrusion foamed using CFC-114 (6 to 8 wt%), into cellular products having density in the 106 to 136 kg/m³ range.

Compatibilization by ionic interaction provided a new way to produce foamable PE/PS blends based on their respective ionomers [39]. Self-associated ionomers are thermally reversible, and thus act as temporary crosslinks [40]. The foamable blends contained a styrene–acrylic acid copolymer (at least 60 wt%) and an ethylene–acrylic acid copolymer (at least 1 wt%). A CBA (alkali metal bicarbonate, 3 to 12 wt%) was used during the extrusion foaming process and the reported foams had a density in the 200 kg/m³ range, with cell sizes of approximately 1 mm.

The choice of an adequate nucleating agent was found to be crucial for the foaming process of polycarbonate resin, and satisfactory results were obtained with core-shell impact modifiers, with preference for acrylate-based multiphase composite interpolymer resins [41]. This patent aimed at finding a foamable PC mixture with good mechanical properties. The addition of fibers was shown to enhance PC foamability but a significant reduction in mechanical properties was observed [42]. In contrast, excellent results were reported using an acrylate impact modifier as the nucleating agent. It seems that these inclusions were highly beneficial to the enhancement of the melt strength properties. Structural foams (ρ = 900 kg/m³) were produced from blends of PC with a polyester resin (PET for example), a minor amount of CBA (up to 2 wt%), a nucleating agent, and a surface active agent as a processing aid.

3.3.2 Nucleation, Growth, and Foam Density

Heterogeneous nucleation is usually preferred in industrial practice since it exerts better control over the cell nucleation process, thus providing structure reproducibility. This is done in many cases using mineral nucleating agents such as talc. However, it has been reported that elastomeric particles can provide a better control of the nucleation stage. Concentration of the rubbery particles dictated the cell nucleation density in a predictive proportionality, with possible incorporation of 0.1 to 18 vol% of either acrylic or di-olefinic elastomer latex particles into a wide variety of polymers (polyester, PE, styrene-acrylonitrile [SAN], PPE/PS, PC, polyetherimide [PEI], PA and fluoropolymer). This approach was developed for microcellular foaming, but is restricted to the solid-state batch foaming process (autoclave foaming). The mechanism controlling the nucleation relies on the voids resulting from stresses induced by the difference in thermal expansion coefficients between the two phases [43].

Foams with densities lower than 20 kg/m³ are classified as ultra-low-density foams, and the huge expansion required may necessitate specific considerations. Ultra-low-density PO foam, having a density in the range of 9.6 to 24 kg/m³, can be extrusion foamed from a blend of PE and/or its copolymer with an elastomer (3 to 30 wt%) and a styrenic polymer or copolymer (1 to 15 wt%) [44]. The PFA used should be a hydrocarbon having 1 to 6 carbon atoms, with isobutane being the preferred candidate. Its concentration may depend on the final foam density desired, and preferably lies between 7 to 40 wt%. This mixture should be combined with a permeability modifier, esters of fatty acids with polyols for example, and a nucleating agent. Examples given in the patent were based on blends of LDPE (70 to 99 wt%), PS (0 to 6 wt%), and SBR (0 to 20 wt%). Densities of the resulting foams as a function of the PFA content are displayed in Figure 3.6.

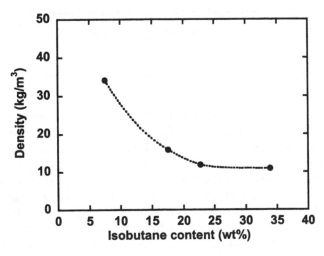

FIGURE 3.6
Foam density reduction with increasing PFA content achieved with an LDPE/PS/SBR blend. (Adapted from Rogers, J.-V. and Kisner, R.D., U.S. Patent 5,225,451, 1993 and U.S. Patent 5,290,822, 1994 [44].)

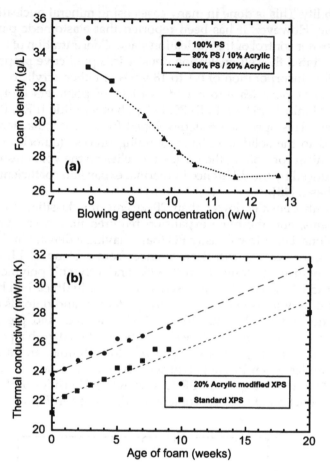

FIGURE 3.7
Blending acrylic with PS improved (a) foam density reduction and (b) insulation properties. (From Smith, P.J. et al., *Proc. 56th ANTEC Conf.*, 1922, 1998 [46]. With permission.)

A significant reduction of foam density can also be accomplished by incorporating an acrylic polymer (up to 50 wt%) to a styrenic polymer [45]. The acrylic polymer can be either a homopolymer or a copolymer, with preference for the latter containing both methacrylate and acrylate moieties. With PS foams being the reference, a density reduction of up to 29% was reported with the incorporation of the acrylic polymer. Some examples are illustrated in Figure 3.7, obtained from extrusion foaming with a PFA mixture of HCFC-22 and HCFC-142b (60/40). It was claimed that PFAs such as hydrocarbons, halogenated hydrocarbons, and inert gases could also be used. In addition, substantial gain in thermal insulation was observed with the added acrylic [46]. Unfortunately, the main drawback from increasing the acrylic polymer content was a reduction in the compressive strength of the foam.

Attempts to extend the usefulness of this blend to extrusion foaming based on HFC-134a or a mixture of HFC-134a and HFC-152a (85/15) has also been

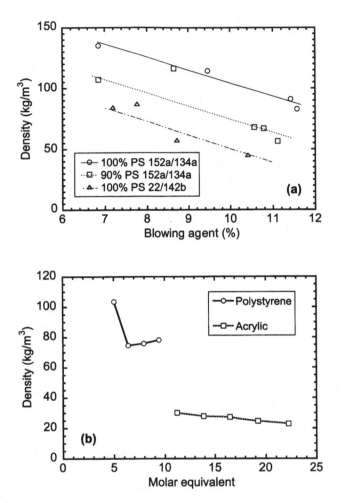

FIGURE 3.8
(a) Impact of blending acrylic with PS on foam density, for PFA mixture of HFC-152a/HFC-134a. (b) Minimum density reached for PS/HFC-134a due to solubility limitations, compared to acrylic foams based on the same PFA. (From Smith, P.J. et al., *Proc. Porous, Cell. Microcell. Materials ASME Conf.*, 82, 99, 1998 [33]. With permission from ASME.)

reported [33]. In addition to the aforementioned foam density reduction (Figure 3.8[a]), incorporation of the acrylic extends the solubility of HFC-134a, which can then be processed at a much higher content than 7.5 wt%, the value usually reported as the limit for PS foaming (Figure 3.8[b]).

3.3.3 Open/Closed Cell Content

While some applications require that all cells should be closed (for example, for insulation purposes), other applications, such as cushioning, necessitate an adequate tuning of the ratio of open/closed cells, which ultimately controls the dynamic of the recovery after compression. Sealing joints in buildings or roads are another application where both open (e.g., 60 to 80%) and closed cells should coexist in the extruded foam. Extruded foams having an adjustable open cell content between 30 and 85% could be produced by blending LDPE with ethylenic copolymers (ethylene–vinyl acetate [EVA], EMA, ethylene–ethyl acrylate [EEA], and ethylene–acrylic acid [EAA]). Blending LDPE with, preferably, EVA (25 to 75 wt%) and foaming the resulting blend using a PFA (CFC-12, CFC-114, HCFC-22, or hydrocarbons) with appropriate additives (nucleating agent, permeability modifier, etc.) yielded the desirable cell structure. While LDPE alone leads to a closed-cell foam, the addition of EVA provides an efficient way to control the level of open-cell content [47]. This in turn affects the recovery time after compression: while an LDPE closed-cell foam compressed to 50% during 7 days did not recover its shape by more than 5% after 1 day, the LDPE/EVA open/closed cell foam got back up to 80% of its original shape after less than 3 hours, with complete recovery after only 1 day.

Fluid transmitting PO foams may have several applications: synthetic leathers, separator of storage battery, filter material, etc. Such open-cell foamed films can be prepared using a PO–styrenic blend [48]. The poor compatibility between the two resins was considered to be the key parameter for the cell opening. Compositions from 1 to 99 wt% were claimed for the styrenic component (PS, PMMA, ABS, with preference for the 1 to 30 wt% range in order to maintain the polyolefin properties) in which a foaming agent, either chemical or physical, was incorporated. The foamable styrenic resin was mixed with the PO (LDPE, HDPE, EVA, LLDPE, PP, or EP) and additional CBAs. The melt blending of the components was achieved during the extrusion foaming. Foams having a density of 500 kg/m^3 and pore sizes of 0.1 mm were obtained from a set composition of 5 to 7 wt% PS, in which 6 wt% of pentane was dissolved.

Enhanced control of the open-cell content was also achieved using ethylene-co-styrene interpolymers (ESIs). These interpolymers are synthesized using single-site, constrained-geometry catalysts, and are remarkable compatibilizers for PE/PS blends due to their inherent combination of both olefinic and styrenic functionalities. Their properties are linked to the styrene copolymer content. Below 45 wt% of styrene copolymer, the ESI is semicrystalline with good low-temperature toughness. Above this limit, it is essentially amorphous with modulus and elongation being functions of the styrene content. Although foams can be made directly from ESI, blending them with various thermoplastics (PE, PP, PS, EVA) has permitted the creation of unique foams. The reported examples included blends of LDPE/ESI and PS/ESI. The former blend yielded lower density foams with smaller cell sizes than its neat LDPE counterpart, while the second blend, compared to a PS foam, was softer, tougher, and more permeable to the isobutane used for foaming. An enhanced control of the open-cell content was achieved through the adequate selection of the styrene copolymer content. Foams made from PS/ESI blends also showed improved toughness and flexibility [49,50].

3.3.4 PFA Affinity

Due to its low cost and environmental friendliness, carbon dioxide is a very attractive foaming agent. Unfortunately for PS extrusion foaming based on CO_2, the expansion is difficult to control since nucleation and growth occurs rapidly, due to the high diffusivity of carbon dioxide. To circumvent these problems, modification of the PS chains was explored in order to induce chemical bonding between the CO_2 and the PS [51]. A styrenic copolymer functionalized with ethylenediamine (EDA) was produced, with the amine groups reacting with CO_2 in a reversible way. Validation was performed with production of microcellular foams exhibiting good mechanical properties. Other CO_2-philic compounds have been highlighted by the same research group; these include fluoroalkyl, fluoroether, and siloxane functional materials [52,53].

Another attempt to modify the solubility and diffusivity of PS through the addition of CO_2-philic additives was performed using aromatic additives containing polar carbonyl groups and poly(dimethyl siloxane) [54]. The latter was blended with PS up to a concentration of 5 wt%, which yielded a slight impact on the solubility of carbon dioxide, increasing it from 9.7 to 10.6 wt% under conditions set at P = 7.58 MPa and T = 35°C. Extrusion foaming using these formulations unfortunately did not indicate any major difference between the foam samples.

3.3.5 Dimensional Stability and Expandable Bead Formulations

Dimensional stability of foams can be very important in many applications. For example, in the lost foam casting process, the final dimension of the thermoplastic patterns molded from expandable beads is critical. The metal casting manufacturing method consists of the following steps. The molded pattern is coated with a refractory material and surrounded by dry sand to form a mold. The molten metal is subsequently poured in that mold, thus vaporizing the foam part and reproducing its shape with accurate details within tight tolerances. Incomplete decomposition of the pattern is undesirable as it creates defects in the metal part [55]. Adequate decomposition and dimensional stability was obtained in one case with a blend obtained from copolymerization of styrene–methacrylic acid ester (55 to 85 wt% of styrene), with the resulting copolymer being impregnated with 10 wt% or less of a physical foaming agent, preferably pentane [56].

Other compositions of expandable styrene polymer-based blends that exhibit reduced shrinkage after foaming and molding were reported by Hahn et al. in 1991 [57]. In this case, styrene polymers (95 to 99.5%) are blended with styrene–acrylonitrile (SAN) copolymer (0.5 to 5 wt%) with the content of acrylonitrile moieties being superior to 0.15 wt% with respect to the total content. The standard PFAs for expandable bead technology (C_3–C_6 hydrocarbon) were used, with content at 3 to 10 wt%. As displayed in Figure 3.9, the degree of shrinkage is continuously reduced with an increasing content of acrylonitrile moieties.

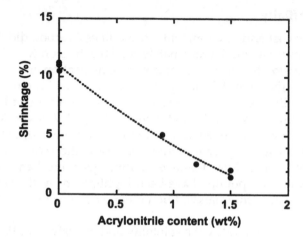

FIGURE 3.9
Reduction of post-molding shrinkage from expandable PS beads incorporating various fractions of SAN. (Adapted from Hahn, K. et al., U.S. Patent 4,983,639, 1991 [56].)

A further enhancement of this blend was claimed the following year [57], where PPE (1 to 20 wt%) was added to improve the heat distortion resistance. The addition of PPE is also highly beneficial for mechanical properties, such as flexural strength. This blend enables the production of foams with much lower densities than conventional PS foams while maintaining similar properties [58].

Expandable bead technology would greatly benefit from an increased number of polymer types that can be foamed using the same steam-chest molding process. Two requirements, however, need to be fulfilled: diffusion of the PFA out of the beads should be slow for adequate shelf life of the expandable particles, and expansion should occur at a temperature of 100°C, as for PS beads. Expandable PP-based particles were prepared according to the following [59]. Emulsion polymerization of styrene (40 to 80 wt%) onto copolymer of diene and propylene resulted in propylene–styrene interpolymer (PSI) that was further blended with a second polymer (PE, ionomers, EVA, EEA, chlorinated PE [CPE], chlorosulfonated PE [CSPE], EPDM, styrene-butadiene block copolymer [SBS], EPR, and ESI). The composition was 75 to 96 wt% of PSI, with 4 to 25 wt% of the other selected polymer. After blending in presence of peroxide (0.005 to 0.5 wt%) to visbreak the PSI, extruding, and pelletizing, the particles were subsequently impregnated with 3 to 20 wt% of a hydrocarbon-based PFA (aliphatic, cycloaliphatic, or halogenated hydrocarbons). The resulting expanded beads led to density in the 25 to 65 kg/m^3 range.

Dimensional stability problems for foam can also originate from the fast permeation of the foaming gas out of the foam matrix, with a relatively slow diffusion of incoming air [60]. Blending with a polymer that has low permeability to gases of interest can be part of the solution, as reported previously for PS–ESI blends [49,50].

3.3.6 Other Physical and Mechanical Properties

Research in blending polyolefins with polystyrene in foam processing is mainly focused on improving foam characteristics. Blending PE with PS should eliminate or at least reduce the drawbacks inherent to each polymer, while combining their respective advantages. For example, polystyrene suffers from brittleness, poor solvent resistance, and a low continuous use temperature (CUT) in the 50 to 70°C range, which limit its use. As a direct consequence, this polymer is not in high demand except for foam applications, because PS and its blends can be foamed easily [4].

Nevertheless, rigid PS foams exhibit poor cushioning capacity. Their inadequacy in absorbing repeated impacts is another major drawback. In addition to their low stiffness, the performance of PE foams is poor in terms of thermal and creep resistance. Cushioning materials should combine flexibility and stiffness, with these properties being stable at a temperature up to 70°C [61]. If insulation properties are targeted, low water permeability as well as some flexibility should also be present. Because PE foams primarily address the cushioning objectives and PS foams address the heat insulation requirements, a balanced combination of both sets of properties would be satisfied by blends of PE and PS. In order to obtain a homogeneous blend that would yield satisfactory performances, constraints on the relative flow properties of the two resins were imposed by Hoki and Miura [61], with a ratio R between 7 and 90. This R variable was defined as the ratio of the melt flow rate (MFR) of PS (1.4 to 18 g/min) to the melt index (MI) of the PE (0.2 to 2.6 g/min). Constraints were also put on the selection of the PFA, whose Kauri-Butanol value [62], which measures the hydrocarbon solvent strength, should fall between 15 and 22. Selected examples excerpted from the patent are reported in Figure 3.10. The foam samples were produced using a concentration of CFC-12 adjusted to yield the desired foam density. Figure 3.10 illustrates that compressive strength and creep properties could be finely tuned by adjusting the foam density and the PS content for these PE/PS blends.

The immiscible blends resulting from blending PS with polyolefins (PO) may also require compatibilization. In order to benefit from low shrinkage and moisture absorption as well as improved impact strength, especially for packaging applications, compatibilizers such as CSPE, styrene–ethylene–butylene–styrene triblock copolymer (SEBS), EVA, or SBS should be added. The numerous benefits listed previously should explain the large number of existing patents on compatibilized PO blends with styrenics. A full range of polymer composition is usually claimed, with emphasis on the need for a compatibilizer having a concentration that can be as high as 10 wt% [4].

As reported by Park et al. [63], compatibilizers for PE/PS blends can be any compound that has molecular segments (ethylene and styrene) compatible with both polymer domains, including the following:

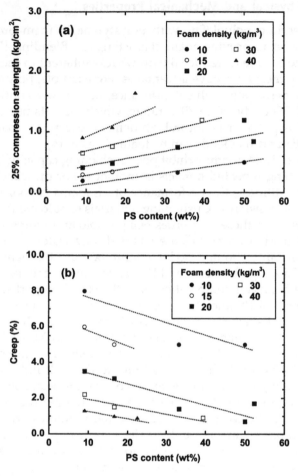

FIGURE 3.10
Variation in the cushioning properties, through adequate selection of PS content and amount of PFA (which impacts the foam density), for foamed PE/PS blends. (Adapted from Hoki, T. and Miura, N., U.S. Patent 4,652,590, 1987 [61].)

- Graft copolymer of PS and PE
- Graft copolymer of PS and EPDM
- Diblock copolymers of styrene–diene
- Hydrogenated diblock copolymers of styrene–diene
- Block copolymers of styrene–ethylene
- Triblock copolymers of styrene–diene–styrene

For example, compatibilization of a blend of PS (5 to 95 wt%) and polyolefin (5 to 95 wt%) was accomplished with a hydrogenated styrene-butadiene block copolymer (SEBS) having a concentration of 0.5 to 10 wt% (in relation to the total mass). Selected foaming examples are listed in Table 3.2, using methyl

TABLE 3.2

Selected Examples of Foams Obtained from Compatibilized PO–PS Blends

PO Type	PO Parts	PS Parts	SEBS Parts	PFA (16 wt%)[b] Composition MeCl (wt%)[c]	C4 (wt%)[c]	Foam Density (kg/m³)	Open-Cell Content (%)	Advantages
LDPE	80	20	3	60	40	40	0	—
LDPE	20	80	3	80	20	26	0	—
HDPE	60	40	5	50	50	30	15	Solvent resistance
EVA[a]	50	50	2	50	50	25	15	Compatible with solvent-base adhesive
EVA	50	50	2	100	0	30	0	Organic solvent resistance
EVA	40	60	3	100	0	20	12	Stable shape (when immersed in butyl acetate)
EVA	40	60	0	100	0	12	12	No shape integrity; disintegrates when immersed in butyl acetate

[a] EVA contains 12 wt% of vinyl acetate.

[b] Weight percentage based on overall composition.

[c] Weight percentage based on PFA content only.

Source: Adapted from Zeitler, G. and Mueller-Tamm, H., U.S. Patent 4,020,025, 1977 [64].

chloride (MeCl) and butane (C4) as the physical foaming agents [64]. As can be noted from the last two examples listed in Table 3.2, compatibilization is essential to property enhancement. However, superior performance was claimed for foams based on PE/PS blends using a specific pseudo-random interpolymer that comprised ethylene and styrene as the compatibilizer [63].

Reactive compatibilization was used in another patent based on LDPE or LLDPE (50 to 95 wt%) blended with PS or HIPS (5 to 50 wt%) [65]. It was suggested that the decomposition of a peroxide (dicumyl peroxide) during the extrusion process induced partial crosslinking of the ethylenic polymer, with the depolymerization of the styrenic polymer occurring simultaneously. The viscosities of the two species were thus altered continuously towards rheological conditions more suitable for foaming. CBA or PFA (halogenated hydrocarbons with 1 to 4 carbons, or hydrocarbons with 3 to 5 carbons) could be used. Reported low-density foams, based on blends of 80 wt% of LDPE and 20 wt% of PS and using CFC-12, showed good dimensional stability. The foams had densities of 25 to 37 kg/m^3 with cell sizes of about 1.5 mm, and open-cell content in the 30 to 70% range. Switching from LDPE to LLDPE kept the foam density constant, reduced the cell size down to 0.1–0.4 mm, and increased the open-cell content to nearly 95%.

Other foamable polymer blends, with their associated advantages related to processability, foam characteristics, and mechanical and physical property enhancements, are listed in Table 3.3. Although this list may appear exhaustive, it is far from being complete, and aims essentially at presenting additional opportunities that can be achieved through blending.

3.4 Conclusions and Outlook

Polymer blending is a well-developed and accepted technology that has achieved large market penetration. The understanding of the critical underlying concepts required for successful polymer blending has greatly matured over the last decade. Rules have been developed for upgrading resin performance by blending processes. Most polymer manufacturers are now proposing high-performance blends based on their own resins, a particularly affordable way to broaden the range of materials and the properties they offer.

Commercial foaming processes are actually based for the most part on a very limited set of thermoplastic polymers: PS, PE, PVC, and to a much smaller extent PP and PET. Most of these foams are essentially produced from neat resins. The most notable exceptions are those based on linear/branched polyolefins blends such as the LDPE/HDPE and PP/HMS-PP systems. These blends are specifically formulated to improve the foamability of the base linear resin. There is then a huge potential for developing a whole range of new materials customized for specific applications, and some exam-

TABLE 3.3

Foamable Polymer Blends and Associated Characteristics

	Formulation		Foam's Characteristics											
Blend Components[a]	Foaming Agent		Crosslinked (I: Irradiation)	Formability (enhanced)	Open (O)/Close (C) Cells	Low Density Foam	Fine Cells	Diffusion Control	Mechanical Properties	Dimensional Stability	Softness (S), Flexible (F)	Heat (H), Solvent (S) Resistance	Comments	Refs.
(PS, SAN) + (diene rubber) + (SMA)	CBA			X			X						Structural foam	67
(PS) + (polybutadiene rubber)	PFA			X									Expandable beads	68
(PB) + (SEBS) + (EPR, EPDM) + (PE)	CBA, PFA							X	X		F	H		69
(PS, SAA) + (vinylidene chloride/methyl acrylate copolymer)	PFA					X		X	X					70
(PE, PP, EP, EVA, or PS) + (PE, PP, EP, EVA, or PS) + (silane-modified base resin)	CBA		X						X	X	S	H	Heat-insulating properties	71
(PO, PS, PMMA, PVDC) + (EPDM)	CBA		X							X				72
(LDPE, LLDPE, HDPE, EVA, EB, PP, EP) + (PS, PMMA, ABS)	HC + CBA				O								Semi-permeable films	48
(LDPE, HDPE, LLDPE, PP, EP) + (PS, sPS, HIPS, PVT, PMS) + (ESI)	CBA			X						X				63
(PE, EVA, PP) + (PS, PMS) + (SEBS)	Halogenated HC, HC					X						S	See Table 3.2	64

(continued)

TABLE 3.3 (CONTINUED)

Foamable Polymer Blends and Associated Characteristics

Formulation		Foam's Characteristics											
Blend Components[a]	Foaming Agent	Crosslinked (I: Irradiation)	Formability (enhanced)	Open (O)/Close (C) Cells	Low Density Foam	Fine Cells	Diffusion Control	Mechanical Properties	Dimensional Stability	Softness (S), Flexible (F)	Heat (H), Solvent (S) Resistance	Comments	Refs.
(PE) + (EMAA ionomer, EVA)	CFC-12/CFC-114				X		X	X	X			Controlled cell growth	73
(LDPE) + (EVA)	CFC, HCFC		X	+	X			X				Open/closed cell in any ratio	74
(EVA) + (PE)	CFC + FC			C	X								75
(EMAA-isobutyl acrylate ionomer) + (EVA-CO) + (elastomer)	CBA	X						X	X				76
(ionic PS) + (ionic PE)	CBA		X					X			S	Foam sheet (blown film extrusion)	39
(PE) + (PE); different molecular weights	Isobutane + PFA	X	X										77
(PE) + (EAA ionomer, PS)			X				X		X				78
(PS polymerized in styrene-swollen LDPE)	HC		X									Expandable beads	79
(LDPE) + (PS)		(X)	X		X				X				65

Blend	Foaming agent							Comments	Ref
(LDPE) + (PS)	CFC			X	X				61
(LDPE) + (HDPE) + (PO wax)	CBA			X	X			Foamed wire insulation	80
(LDPE) + (HDPE) + (PP) + (silane coupling agent)	CBA, inert gases	(X)		X	X			Coaxial cable	81
(LDPE) + (diene elastomer) + (PS)	HC	X		X				Ultra-low density	44
(PE) + (PE grafted with silyl groups)	PFA	X		X	X				82
(crystalline propylene copolymer) + (MDPE)	CBA, PFA	I	C	X	X				83
(MDPE) + (LDPE)	CFC		C	X					84
(LDPE) + (LLDPE)	PFA (CFC)			X		F	H		85
(LDPE) + (LLDPE)		X			X				86
((EPR, EPDM) + (PP) + (PIB, IIR)) + (LLDPE)	CBA	(X)		X		F			87
(PE) + (PIB), (PP) + (EPR)	CFC	X	C	X					88
(PSI) + (PE, ionomers, EVA, EEA, CPE, CSPE, EPDM, SBS, EPR, ESI)	HC, halogenated HC			X	X			Expandable beads	59
((EPDM, EPR, IIR) + (PP)) + (bPP)	CBA					S	H		89
(EPR, EPDM) + (EVA)	CBA	I					H		90
(EAA, EMAA) + (EVA, EEA)	PFA	X	C	X					91

[a] For polymer abbreviations, see list on pages 134–136.

Source: Adapted from Utracki, L.A. *Polymer Blends Handbook*, Klewer Academic Publ., 2002 [5] and Utracki, L.A. et al., NRC Report, 1999 [66].

ples have been briefly reviewed in this chapter. As foam users are specifying more and more stringent requirements, polymer blending can be seen as an obvious solution for producing foam with improved performance.

As mentioned in the introduction to this chapter, the literature on polymer blend foaming is rare. In addition to the commercial opportunities offered by this topic, polymer blending remains a formidable scientific challenge addressed to the research and development community involved in polymer foam development.

Abbreviations

ABS:	Acrylonitrile–butadiene–styrene copolymer
aPP:	Atactic polypropylene
bPP:	Branched polypropylene
C4:	Butane
CBA:	Chemical blowing agent
CFC:	Chlorofluorocarbon
coPA:	Copolyetheresteramide
CPE:	Chlorinated PE
CSPE:	Chlorosulfonated polyethylene
CTM:	Cavity transfer mixer
CUT:	Continuous use temperature
DVB:	Divinyl benzene
EAA:	Ethylene–acrylic acid copolymer
EB:	Ethylene–butene copolymer
EDA:	Ethylenediamine
EEA:	Elastomeric copolymer from ethylene and ethyl acrylate
E-EA–GMA:	Elastomeric terpolymer from ethylene, ethyl acrylate, and glycidyl methacrylate
EFM:	Extensional flow mixer
EGMA:	Copolymer from ethylene and glycidyl methacrylate
EMA:	Copolymer from ethylene and maleic anhydride or ethylene and methyl acrylate
EMAA:	Ethylene–methacrylic acid copolymer
EP:	Ethylene–propylene copolymer
EPDM:	Elastomeric terpolymer from ethylene, propylene, and a nonconjugated diene

EO:	Elastomeric copolymer from ethylene and octene
EPR:	Elastomeric copolymer of ethylene and propylene
ESI:	Ethylene–costyrene interpolymer
EVA:	Ethylene–vinyl acetate copolymer
EVA-GMA:	Elastomeric terpolymer from ethylene, vinyl acetate, and glycidyl methacrylate
EVOH:	Ethylene–vinyl alcohol copolymer
FC:	Fluorocarbon
GMA:	Glycidyl methacrylate
HC:	Hydrocarbon
HCFC:	Hydrochlorofluorocarbon
HDPE:	High-density polyethylene
HDT:	Heat deflection temperature
HFC:	Hydrofluorocarbon
HIPS:	High-impact polystyrene
IIR:	Isoprene–isobutylene rubber; butyl rubber
LCP:	Liquid crystalline polymer
LDPE:	Low-density polyethylene
LLDPE:	Linear low-density polyethylene
MDPE:	Medium density polyethylene
MeCl:	Methyl chloride
MFR:	Melt flow rate
MI:	Melt index
MW:	Molecular weight
MWD:	Molecular weight distribution
NBR:	Elastomeric copolymer from butadiene and acrylonitrile
PA:	Polyamide
PB:	Polybutylene
PBT:	Polybutylene terephthalate
PC:	Polycarbonate
PCL:	Poly(ε-caprolactone)
PE:	Polyethylene
PEI:	Polyetherimide
PEO:	Poly(ethylene oxide)
PET:	Polyethylene terephthalate
PFA:	Physical foaming agent
PIB:	Polyisobutene

PMMA:	Poly(methyl methacrylate)
PMS:	Poly(α-methylstyrene)
PO:	Polyolefin
PP:	Polypropylene
PP-HMS:	High melt strength polypropylene; a marketing term for branched polypropylene
PPE:	Poly(2,6-dimethyl 1,4-phenylene ether)
PS:	Polystyrene
PSI:	Propylene–styrene interpolymer
PVC:	Poly(vinyl chloride)
PVDC:	Poly(vinylidene chloride)
PVT:	Poly(vinyl toluene)
SAA:	Styrene–acrylic acid copolymer
SAN:	Styrene–acrylonitrile copolymer
SCF:	Supercritical fluid
SBS:	Styrene–butadiene–styrene triblock copolymer
SEBS:	Styrene–ethylene–butylene–styrene triblock copolymer
SMA:	Styrene–maleic anhydride copolymer
sPS:	Syndiotactic polystyrene
SSE:	Single-screw extruder
TPS:	Thermoplastic starch
TPU:	Thermoplastic poly(urethane)
TSE:	Twin-screw extruder

References

1. Utracki, L.A., *Polymer Alloys and Blends*, Hanser, Munich, 1989.
2. Utracki, L.A., Ed., *Polymer Blends Handbook*, Kluwer Academic Publishers, Netherlands, 2002.
3. Utracki, L.A., Polymer Blends, *Rapra review reports*, 11(3), 1, 2000.
4. Utracki, L.A., *Foaming Polymer Blends*, NRC report #43033, 1998.
5. Utracki, L.A., Introduction to polymer blends, in *Polymer Blends Handbook*, Kluwer Academic Publishers, Netherlands, 2002, chap. 1.
6. Olabisi, O., Robeson, L.M., and Shaw, M.T., *Polymer–Polymer Miscibility*, Academic Press, New York, 1979.
7. Wu, S., A generalized criterion for rubber toughening: the critical matrix ligament thickness, *J. Appl. Polym. Sci.*, 35, 549, 1988.
8. Chapleau, N. and Huneault, M.A., Impact modification of poly(ethylene terephthalate), *J. App. Polym. Sci.*, 90, 2919, 2003.

9. Isayev, A.I. and Viswanathan, A., Self-reinforced prepregs and laminates of a poly(phenylene oxide)-polystyrene alloy with a liquid crystalline polymer, *Polymer*, 36, 1585, 1995.
10. Champagne, M.F. et al., Generation of fibrillar morphology in blends of block copolyetheresteramide and liquid crystal polyester, *Polym. Eng. Sci*, 36, 1636, 1996.
11. Champagne, M.F., Huneault, M.A., and Roberge, H., Blend morphology of HDPE/PA6 blow molded bottles, *Proc. 53rd ANTEC Conf.*, 1552, 1995.
12. Subramanian, P.M. and Mehra, V., Laminar morphology in polymer blends: structure and properties, *Polym. Eng. Sci.*, 27, 663, 1987.
13. Kamal, M.R. et al., The development of laminar morphology during extrusion of polymer blends, *Polym. Eng. Sci.*, 35, 41, 1995.
14. Holsti-Miettinen, R.M. et al., Oxygen barrier properties of polypropylene/polyamide 6 blends, *J. Appl. Polym. Sci.*, 58, 1551, 1995.
15. Champagne, M.F. et al., Reactive compatibilization of polypropylene/polyethylene terephthalate blends, *Polym. Eng. Sci.*, 39, 976, 1999.
16. Rodriguez-Gonzalez, F.J., Ramsay, B.A., and Favis, B.D., High performance LDPE/thermoplastic starch blends: a sustainable alternative to pure polyethylene, *Polymer*, 44, 1517, 2003.
17. Seidenstucker, T. and Fritz, H.-G., Innovative biodegradable materials based upon starch and thermoplastic poly(ester-urethane) (TPU), *Polym. Deg. Stab.*, 59, 279, 1998.
18. Tattum, S.B., Cole, D., and Wilkinson, A.N., Controlled transesterification and its effects on structure development in polycarbonate-poly(butylene terephthalate) melt blends, *J. Macromol. Sci. Part B: Phys.*, 39, 459, 2000.
19. Ma, D.Z. et al., Compatibilizing effect of transesterification product between components in bisphenol-A polycarbonate/poly(ethylene terephthalate) blend, *J. Polym. Sci. Part B: Polym. Phys. Ed.*, 37, 2960, 1999.
20. Gale, G.M., Use of the cavity transfer mixer in compounding, *Plast. Comp.*, 70, May/June 1985.
21. Luciani, A. and Utracki, L.A., The extensional flow mixer, EFM, *Intern. Polym. Process.*, 11, 299, 1996.
22. Datta, S. and Lohse, D.J., *Polymeric Compatibilizers: Uses and Benefits in Polymer Blends*, Hanser, New York, 1996.
23. Lohse, D.J., Russell, T.P., and Sperling, L.H., Eds., *Interfacial Aspects of Multicomponent Polymer Materials*, Plenum Press, New York, 1997.
24. Elemans, P.H.M., Janssen, J.M.H., and Meijer, H.E.H., The measurement of interfacial tension in polymer/polymer systems: the breaking thread method, *J. Rheol.*, 34, 1311, 1990.
25. Luciani, A., Champagne, M.F., and Utracki, L.A., Interfacial tension in polymer blends. Part 2: Measurements, *Polym. Netw. Blends*, 6, 51, 1996.
26. Lin, B. et al., Erosion and breakup of polymer drops under simple shear in high viscosity ratio systems, *Polym. Eng. Sci.*, 43, 891, 2003.
27. Huneault, M.A. et al., Dispersion in high viscosity ratio polyolefin blends, *Polym. Eng. Sci.*, 41, 672, 2001.
28. Lin, B. et al., Parallel breakup of polymer drops under simple shear, *Macromol. Rapid Comm.*, 24, 783, 2003.
29. Elkovitch, M.D., Lee, L.J., and Tomasko, D.L., *Polym. Eng. Sci.*, 40, 1850, 2000.
30. Lee, M., Tzoganakis, C., and Park, C.B., Extrusion of PE/PS blends with supercritical carbon dioxide, *Polym. Eng. Sci.*, 38, 1112, 1998.

31. Pesneau, I. et al., Foam extrusion of PP-EMA reactive blends, *J. Cell. Plast.*, 38, 421, 2002.
32. Utracki, L.A. and Kamal, M.R., The rheology of polymer alloys and blends, in *Polymer Blends Handbook*, Utracki, L.A., Ed., Kluwer Academic Publishers, Netherlands, 2002, chap. 7.
33. Smith, P.J. et al., Polystyrene/acrylic blends and their application in foam production. Part two: R134a blowing agent, *Proc. Porous, Cell. Microcell. Mater. ASME Conf.*, MD-Vol. 82, 99, 1998.
34. McCracken, W.J., Nitrile/PVC and other polymer/resin closed cell foams, *J. Cell. Plast.*, 20, 150, 1984.
35. Allen, R.B. et al., U.S. Patent 4,857,390, 1989.
36. Van Calster, M., High melt strength PP foams for the automotive and packaging industries, *Proceedings of the Foamplas '97 Conference*, Mainz, Germany, 1997, 151.
37. Van Nuffel, C.T.E. et al., U.S. Patent 5,804,673, 1998.
38. DeNicola, A.J., Jr., Smith, J.A., and Felloni, M., U.S. Patent 5,605,936, 1997; U.S. Patent 5,414,027, 1995.
39. Park, C.P., U.S. Patent 4,567,209, 1986; U.S. Patent 4,623,671, 1986; and U.S. Patent 4,722,972, 1988.
40. Brown, S.B., Reactive compatibilization of polymer blends, in *Polymer Blends Handbook*, Utracki, L.A., Ed., Kluwer Academic Publishers, Netherlands, 2002, chap. 5.
41. Avakian, R.W. and Jodice, R.E., U.S. Patent 4,587,272, 1986.
42. Fox, D.W., U.S. Patent 4,280,005, 1981; U.S. Patent 4,351,911, 1982.
43. Campbell, G.A. and Rasmussen, D.H., U.S. Patent, 5,358,675, 1994; U.S. Patent 5,369,135, 1994.
44. Rogers, J.-V. and Kisner, R. D., U.S. Patent 5,225,451, 1993; U.S. Patent, 5,290,822, 1994.
45. Smith, P.J. and Cross, B.J., U.S. Patent 6,063,485, 1996.
46. Smith, P.J. et al., Polystyrene/acrylic blends and their application in foam production. Part one: R22/142b blowing agent, *Proc. 56th ANTEC Conf.*, 1922, 1998.
47. Hovis, E.E., Johnson, E.D., and Schroeder, M.J., U.S. Patent 4,931,484, 1990; U.S. Patent 5,059,631, 1991.
48. Tashiro, H. et al., U.S. Patent 4,384,032, 1983.
49. Chaudhary, B.I., Barry, R.P., and Tusim, M.H., Foams made from blends of ethylene styrene interpolymers with polyethylene, polypropylene and polystyrene, *J. Cell. Plast.*, 36, 397, 2000.
50. Diehl, C.F. et al., Blends of ethylene/styrene interpolymers and other polymers: benefits in applications, *Proc. 57th ANTEC Conf.*, 2149, 1999.
51. Diaf, A., Enick, R.M., and Beckman, E.J., Molecular redesign of expanded polystyrene to allow use of carbon dioxide as a foaming agent. I. Reversible binding of CO_2, *J. Appl. Polym. Sci.*, 50, 835, 1993.
52. Shi, C. et al., The gelation of CO_2: a sustainable route to the creation of microcellular materials, *Science*, 286, 1540, 1999.
53. Sarbu, T., Styranec, T.J., and Beckman, E.J., Design and synthesis of low cost, sustainable CO_2-philes, *Ind. Eng. Chem. Res.*, 39, 4678, 2000.
54. Vachon, C. and Tatibouët, J., Effect of additives on the solubility and diffusivity of CO_2 in polystyrene, *Proc. 62nd ANTEC Conf.*, 2004.
55. Brown, J.R., Lost foam casting, in *Foseco Ferrous Foundryman's Handbook*, Butterworth-Heinemann, Oxford, UK, 2000, chap. 15.

56. Kato, Y. et al., U.S. Patent 5,403,866, 1995.
57. Hahn, K., Guhr, U., and Hintz, H., U.S. Patent 4,983,639, 1991; Hahn, K. et al., U.S. Patent 5,093,375, 1992.
58. Koetzing, P. and Diebold, K., EPS/PPE Partikelschaum, *Kunststuffe*, 85, 2046, 1995.
59. Fudge, K.D., U.S. Patent 4,666,946, 1987.
60. Yang, C.-T., Lee, K.L., and Lee, S.-T., Dimensional stability of LDPE foams: modeling and experiments, *J. Cell. Plas.*, 38, 113, 2002.
61. Hoki, T. and Miura, N., U.S. Patent 4,652,590, 1987.
62. American Society for Testing and Materials, ASTM D-1133-02, Standard Test Method for Kauri-Butanol Value of Hydrocarbon Solvents.
63. Park, C.P. et al., U.S. Patent 5,460,818, 1995.
64. Zeitler, G. and Mueller-Tamm, H., U.S. Patent 4,020,025, 1977.
65. Park, C.P., U.S. Patent 4,605,682; U.S. Patent 4,652,588, 1987.
66. Utracki, L.A., Daigneault, L.E., and Moulinié, P., Polyethylene foaming review, NRC report, 1999.
67. Sprenkle, W.E., Jr., U.S. Patent 4,207,402, 1980.
68. Henn, R. et al., U.S. Patent 5,525,636, 1996.
69. Hwo, C.C., U.S. Patent 5,585,411, 1996.
70. Romesberg, F.E., U.S. Patent 5,051,452, 1991.
71. Kobayashi, T., Miyazaki, K., and Nakamura, M., U.S. Patent 5,594,038, 1997; U.S. Patent 5,646,194, 1997.
72. Cakmak, M. and Dutt, A., U.S. Patent 5,114,987, 1992.
73. Ehrenfruend, H.A., U.S. Patent 4,110,269, 1978.
74. Hovis, E.E., Johnson, E.D., and Schroeder, M.J., U.S. Patent 4,931,484, 1990; U.S. Patent 5,059,631, 1991.
75. Park, C.P., Corbett, J.M., and Griffin, W.H., U.S. Patent 4,129,530, 1978.
76. Enderle, S.J., U.S. Patent 4,480,054, 1984.
77. Cree, S.H., Wilson, C.A., and de Vries, S.A., U.S. Patent 5,795,941, 1998.
78. Park, C.P., U.S. Patent 4,640,933, 1987; U.S. Patent 4,663,361, 1987; and U.S. Patent 4,694,027, 1987.
79. Kitamori, Y., U.S. Patent 4,168,353, 1979.
80. Sakamoto, T., Kumai, K.M and Gotoh, S., U.S. Patent 5,346,926, 1994.
81. Sakamoto, T., Inoue, T., and Yoshida, M., U.S. Patent 5,643,969, 1997.
82. Fukumura, M. et al., U.S. Patent 4,456,704, 1984.
83. Nojiri, A. et al., U.S. Patent 4,421,867, 1983.
84. Park, C.P. and Bouton, R.A., U.S. Patent 4,226,946, 1980.
85. Haselier, F.J.J., U.S. Patent 4,939,181, 1990.
86. Nakamura, H. and Iwano, S., U.S. Patent 4,649,001, 1987.
87. Matsuda, A., Shimizu, S., and Abe, S., U.S. Patent 4,212,787, 1980; U.S. Patent 4,247,652, 1981.
88. Searl, A.H., Hahn, G.J., and Rutledge, R.N., U.S. Patent 4,451,586, 1984.
89. Okada, K., Karaiwa, M., and Uchiyama, A., U.S. Patent 5,728,744, 1998; U.S. Patent 5,786,403, 1998.
90. Kagawa, S., Toda, H., and Nomura, S., U.S. Patent 5,057,252, 1991.
91. Park, C.P., U.S. Patent 4,215,202, 1980.

4

Research on Alternative Blowing Agents

Caroline Vachon

CONTENTS

4.1 Introduction

Motivation for the development of alternative blowing agents has been driven by intense political and environmental pressure to stop the destruction of the ozone layer. It was clear from early reports from the Scientific Assessment of Ozone Depletion that major climactic changes in the Arctic were due to the presence of chlorine and bromine in the stratospheric layer. These elements come from the degradation of chlorofluorocarbons (CFCs), compounds that have been extensively used in the foam industry. In 1990, building and appliance insulation applications accounted for approximately 140,000 metric tons of CFCs, while the remaining 34,000 tons were used as auxiliary blowing agents [1]. Common blowing agents included CFC-11, CFC-12, CFC-113, and CFC-114.

There have been numerous concerns about replacing CFCs with alternative compounds, such as toxicity issues, flammability, thermal insulation, and other performance-related properties. The purpose of this chapter is to investigate the latter, which are critical to foam processing. A review of various alternatives will be presented and further evaluated for foam extrusion using well-established characterization techniques. This will provide readers with a global understanding of technical challenges and envisaged solutions for successful foam processing.

4.1.1 Historical Perspective

The quest for safe and efficient refrigeration was at the root of early blowing agent development. At the beginning of the twentieth century, highly toxic and corrosive chemicals (ammonia, butane, methyl chloride [or bromide], sulfur dioxide) were typically used as refrigerants. Problems with safety exerted a constant pressure on the industry to develop alternative compounds. In 1928, Frigidaire commissioned a research team, led by Thomas Midgley, Jr., at General Motors, to find the ideal refrigerant. Midgley and his collaborators proposed that the ideal refrigerant could be formed from halogenated derivatives of aliphatic hydrocarbons. Such compounds were expected to have suitable properties such as low boiling points, nonflammability, and nontoxicity. A first patent was filed in 1930, in which a process of refrigeration was claimed using a series of chlorinated or fluorinated compounds [2], followed by a second application, describing the synthesis of the first CFC from the fluorination of carbon tetrachloride [3]. From the very start, it was obvious that these compounds were remarkably nontoxic. Following their introduction, the use of CFCs experienced dramatic growth in numerous industrial applications such as propellants, foam blowing agents, fire extinguishers (halons), air conditioning, cleaning solvents, and inhalers.

The early development of cellular plastics goes back to the beginning of the twentieth century. The first commercial sponge rubber was introduced between 1910 and 1920 [4], followed by foamed polystyrene in 1931 [5] and foamed polyethylene in 1941 [6]. Extruded and expanded polystyrene were later introduced in 1944 and 1952, respectively [4]. The use of CFCs as foam blowing agents and in other applications would remain unchallenged for many decades, until 1974, when Molina and Rowland published results of a ground-breaking work demonstrating that CFCs present in the atmosphere destroy the ozone layer in an irreversible manner [7]. The ozone contained in the stratospheric layer of the Earth's atmosphere is important because it serves as a protecting barrier from harmful ultraviolet (UV) radiation. Ozone is made of three oxygen atoms (O_3) that absorb UV radiation, preventing it from ever reaching the surface; this radiation could cause skin cancer, reduce the efficiency of the human body's immune system, and upset marine and terrestrial ecosystems [8].

CFCs are extremely stable and can migrate in the atmosphere, where they are broken down by solar radiation, thus releasing chlorine radicals. The latter set off chain reactions that destroy the ozone layer as follows:

$$Cl + O_3 \rightarrow ClO + O_2 \qquad (4.1)$$

$$ClO + O \rightarrow Cl + O_2 \qquad (4.2)$$

As demonstrated in the equations above, the degradation mechanism of ozone is catalytic, as chlorine radicals are regenerated continuously and can participate in as many as 100,000 similar reactions before being finally eliminated from the stratospheric layer. Substances that are fully halogenated have the highest potential for damage; these include CFCs and halons. Substances that contain hydrogen, such as hydrochlorofluorocarbons (HCFCs), are less stable than CFCs, which reduces their persistence in the atmosphere, rendering them less damaging to the ozone layer. For the purpose of comparing the destructive potential of these chemical compounds, an ozone-depleting potential (ODP) was assigned in comparison with CFC-11. The ODP of CFC-11 was arbitrarily set to 1. Table 4.1 lists the ODP and global warming potential (GWP), set to 1 for carbon dioxide, of common CFCs and HCFCs.

In order to stop the destruction of the ozone layer, the Montreal Protocol was developed by the United Nations and signed by more than 100 participating countries around the world. As of January 1, 1989, measures were put forward to decrease and eventually cease the production and use of CFCs. Originally, the agreement was aimed at five CFCs and three halons (brominated compounds used as fire extinguishing agents). However, because other substances such as HCFCs can also deplete the ozone layer, the agreement was extended to other classes of chemicals [8,9].

Generally, the replacement of CFCs used as blowing agents followed two distinct paths. Foams made for noninsulating purposes such as packaging converted from CFCs directly to hydrocarbons (HCs). CFC-114 was widely used for polyolefins, and CFC-12 for polystyrene. Low molecular weight hydrocarbons, namely isomers of butane and pentane, are now processed

TABLE 4.1

Physical Properties of Common Chlorofluorocarbons (CFCs) and Hydrochlorofluorocarbons (HCFCs) [1–3]

Name	Formula	ODP	$GWP_{100\ yr}$	Vapor Pressure at 25°C (kPa)	Boiling Point (°C)	Vapor Flame Limit (vol% air)	Vapor Thermal Conductivity at 25°C (mW/mK)	Ref.
CFC-11	$CFCl_3$	1.0	4,600	—	23.7	none	8.7	10, 11
CFC-12	CF_2Cl_2	1.0	10,600	652	−29.8	none	10.4	10, 11
CFC-113	$C_2F_3Cl_3$	0.8	6,000	—	47.7	none	7.8	10
CFC-114	$C_2F_4Cl_2$	1.0	9,800	213	3.5	none	11.2	10, 11
CFC-115	C_2F_5Cl	0.6	7,200	911	−39.1	none	13.9	10, 11
HCFC-22	CHF_2Cl	0.055	1,700	1040	−40.7	none	11.8	10, 11
HCFC-123	CF_3CHCl_2	0.02–0.06	120	—	27.9	none	10.4	10
HCFC-124	CF_3CHClF	0.02–0.04	620	—	−12	—	—	10
HCFC-141b	CH_3CFCl_2	0.11	700	—	32	7.6–17.7	10.0	10
HCFC-142b	CH_3CF_2Cl	0.065	2,400	339	−9.1	6.7–14.9	9.4	10, 11
HCFC-225ca	$CF_3CF_2CHCl_2$	0.025	180	—	—	—	—	—
HCFC-225cb	$CClF_2CF_2CHClF$	0.033	620	—	—	—	—	—

either as single agents or in blends. Because of technical considerations based on gas diffusivity and resin property, polyolefins are generally processed with butanes (n-butane, isobutane) while polystyrene is processed with pentanes (n-pentane, isopentane, cyclopentane). For insulating foams (polyurethanes and polystyrene), the industry has moved towards HCFCs, which are being used as a transitory replacement until newer compounds can be developed and commercialized. CFC-11 was the blowing agent of choice for polyurethanes. Common replacements include HCFC-141b and HCFC-22. The latter is also used in polystyrene foam as a replacement for CFC-12 as is HCFC-142b. Nowadays, hydrofluorocarbons (HFCs) and other fluorinated derivatives are expected to gradually replace HCFCs for thermal insulating applications.

4.1.2 Nomenclature

Blowing agents (BAs), excluding halons, use a basic nomenclature that describes the number and type of atoms contained in the molecule. The key to the code is to add 90 to the number of the BA, which reveals the number of carbon, hydrogen, and fluorine atoms. For instance, CFC-12 (12 + 90 = 102) contains 1 C, 0 H, and 2 F. The total number of available sites on the carbon is four covalent bonds. Any vacant position on the molecule is filled by chlorine atoms (two in this case). Note that molecules containing one carbon have two-digit numbers, while those containing two or three carbon atoms have three-digit numbers. In some cases, it will become necessary to differentiate isomers. For instance, HFC-134 (134 + 90 = 224) contains 2 C, 2 H, and 4 F. In this particular case, all available bonds are filled by hydrogen or fluorine atoms. However, the fluorine atoms may be distributed evenly or unevenly on the carbon backbone so that the two possible isomers would be 1,1,2,2-tetrafluoroethane or 1,1,1,2-tetrafluoroethane. Isomers are named so that the structure with the atomic weights of the atoms bonded to the carbon atoms most evenly distributed has no letter. 1,1,2,2-Tetrafluoroethane has an even distribution of fluorine atoms on the carbon atoms, so that it is named HFC-134. 1,1,1,2-Tetrafluoroethane is named HFC-134a.

Molecules with three carbon atoms are more complicated to name. They use two letters to designate the atoms attached to the carbon atoms. The first letter designates the atoms attached to the middle carbon, and the second letter the decreasing symmetry in atomic weights of the atoms attached to the outside carbon atoms. The coding describing the atoms on the middle carbon is listed in Table 4.2. HCFC-225ca (225 + 90 = 315) contains 3 C, 1 H, and 5 F. There are two vacant positions filled by chlorine atoms. The letter c indicates that two fluorine atoms are attached to the middle carbon, and the letter a indicates the distribution of the weight of atoms on the outside carbon atoms. Thus, HCFC-225ca has the structure CF_3-CF_2-$CHCl_2$ and isomer HCFC-225cb is $CClF_2$-CF_2-$CHClF$. Miscellaneous classification also exists for inert gases and hydrocarbons. Readers should refer to the standard designation and safety classification of refrigerants for further details [12].

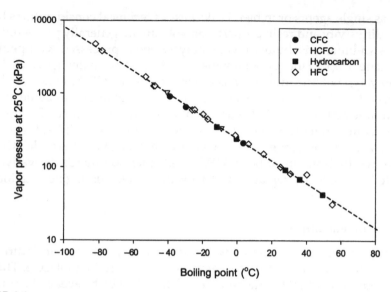

FIGURE 4.1
Correlation between vapor pressure (at 25°C) and boiling point for physical blowing agents.
Data from Tables 4.1, 4.3, and 4.4.

TABLE 4.2

ASHRAE Standard 34-2001 Coding for Describing
Isomers in Three Carbon Molecules

Atoms on Middle Carbon	Code Letter
Cl_2	a
Cl, F	b
F_2	c
Cl, H	d
H, F	e
H_2	f

4.1.3 General Requirements for Physical Foaming Agents

The choice of a suitable alternative physical foaming agent (PFA) must rely
on certain characteristics: adequate vapor pressure, sufficient solubility, low
diffusivity, chemical inertness, safety, and reasonable cost. Vapor pressure is
closely related to the boiling point of a substance (Figure 4.1). Common PFAs
can have boiling points ranging from –50°C to 40°C, and their blowing power
will decrease with increasing boiling point. A good balance between solubility and diffusivity is mandatory for producing suitable foams, as density
can be globally correlated with BA concentration. On one hand, increasing
the BA concentration will decrease the density of the final foam; therefore,
sufficient amounts of PFA must be dissolved in the polymer melt. On the

other hand, PFA solubility should not be too important, since it may over-plasticize the matrix and tend to remain dissolved in the polymer walls after processing, which could be detrimental to mechanical properties. Similarly, diffusion is equally important during both processing and aging of the foam. Should diffusion of the blowing agent become too important, poor cell structure during foaming may result (open cells, coalescence, etc.), or post-processing shrinkage may be generated due to the difference between the rate of permeation of the incoming air vs. the outcoming blowing agent. For instance, n-pentane is a hydrocarbon commonly used for extruded poly-styrene. It has a boiling point of 36°C and can create low-pressure zones in the foam cells upon cooling. Because polystyrene (PS) is rigid, the cells can support the reduced internal pressure without collapsing. This is not the case for LDPE, and important shrinkage will be noted in LDPE foams extruded with n-pentane [13].

Of course, chemical inertness is also required from PFAs, as they should not modify or degrade the processed polymer or be corrosive to the machin-ery. Ideally, PFAs should be as safe and as nonreactive as possible. However, there are certain risks associated with flammable gases such as HCs that can be managed through adequate safety measures. Finally, cost will very often determine the feasibility of using a PFA for production purposes. Producers will use low-cost PFAs for the production of commodity foams such as food trays. Specialty foams, such as insulation panels, however, are made from more costly HCFCs or HFCs, since thermal performances are required.

4.2 Physical Blowing Agents

Physical blowing agents are generally comprised of low-boiling volatile liquids such as aromatic, cyclic, aliphatic, or halogenated hydrocarbons, ethers, ketones, and alcohols that transform into gases at extrusion condi-tions. These gases impart the blowing power to expand the polymer into foam. Permanent gases that can be compressed or directly injected in the extruder can also be used as physical blowing agents. Carbon dioxide and nitrogen remain the two most commonly used permanent gases. Other gases that might be of some use include air, argon, helium, or other noble gases, although cost might limit their actual industrial potential. Because air con-tains oxygen, it is not being used for polymer processing, since the oxidative power of oxygen would certainly degrade the polymer matrix [10].

The blowing power of a gas is largely dictated by its vapor pressure and the volume of gas generated after decompression will limit the minimum reachable density. Ideally, the maximum expansion ratio may be estimated as follows [14]:

Ideal expansion ratio: $\dfrac{Volume_{polymer} + Volume_{gas}}{Volume_{polymer}} = 1 + \dfrac{m_{gas}}{m_{polymer}} \times \dfrac{v_{gas}}{v_{polymer}}$ (4.3)

where m and v are the weight and specific volume, respectively.

Experimentally, the maximum expansion ratio is never reached, since not all of the blowing agent will be efficiently used to make the foam. Gas may partly escape from the polymer skin or remain dissolved in the polymer walls for long periods of time. Moreover, bubble expansion is also going to be limited by the viscosity of the polymer matrix.

General requirements for the efficient expansion of polymer foams include the following [14,15]:

- Plasticization of the polymer matrix
- Complete dissolution of the blowing agent
- Reduction of gas diffusion
- Suppression of cell coalescence
- Promotion of large expansion

Plasticization of the polymer matrix readily occurs through the dissolution of the blowing agent. This can be promoted by using a suitable screw configuration and additional mixing devices, thus helping to prevent the formation of larger cells. Reduction of gas diffusion can be achieved by reducing the polymer melt temperature, also contributing to the improvement of polymer melt strength and the prevention of coalescence. In the case of semicrystalline polymers, the processing window leading to high-volume expansion is very narrow because of the crystallization temperature (T_c). For these polymers, the optimal foam growth is generally obtained only a few degrees above T_c. Moreover, polymer structure will dictate the resistance of the melt to deformation (i.e., melt strength). It is widely recognized that branched semicrystalline polymers offer a melt strength superior to that of linear polymers, owing to their strain hardening behavior [15,16]. Finally, preventing gas escape from the polymer skin can be accomplished by quenching the surface with a low die temperature.

4.2.1 Hydrocarbons

HCs are largely used in the industry to produce low-density commodity foams at moderate cost. Table 4.3 lists current HCs used as PFAs. Rigid polymers such as polystyrene are typically extruded with isomers of pentane while polyolefins are mostly processed with isobutane. The difference lies in the permeability of the blowing agents, which can cause shrinkage in the case of polyethylene. Before the ban of CFCs, low-density polyethylene (LDPE) was generally processed with CFC-114. It had a relative permeability

TABLE 4.3

Common Hydrocarbons Used for Extruded Thermoplastic Foams

Name	Formula	GWP	Vapor Pressure (kPa) at 25°C	Boiling Point (°C)	Vapor Flame Limits (% vol)	Vapor Thermal Conductivity at 25°C (mW/mK)
n-Butane	C_4H_{10}	—	243.6	−0.45	1.9–8.5	15.9
Isobutane	C_4H_{10}	—	351.7	−11.70	1.8–8.4	16.3
n-Pentane	C_5H_{12}	11	68.4	36	1.3–8.0	13.7
Isopentane	C_5H_{12}	11	91.8	27.8	1.4–7.6	13.3
Cyclopentane	C_5H_{10}	11	42.4	49.3	1.4–9.4	11.2

close to that of air, which made it the ideal blowing agent for polyethylene (PE) [17]. Unfortunately, HCs permeate much faster than air, resulting in shrinkage due to the internal vacuum. In order to create dimensionally stable foams, permeability modifiers have been used to retard the diffusion of hydrocarbons. Thus, LDPE foams extruded with isobutane and less than 2 wt% stearyl stearamide were found to be stable [17].

Processing HCs requires special safety provisions in order to deal with the very flammable nature of these compounds. Detectors located near the processing line provide a constant monitoring of the level of HCs in air. Adequate ventilation is also mandatory around the processing site and in the storage area. Aside from these general safety issues, HCs have a moderate or low volatility and are very soluble in most polymers. Thus, processing HCs in polymer melts is fairly straightforward. Plasticization of n-pentane in polystyrene and polyolefins was reported [18] and results show that it is a good plasticizer for both polymers. On a weight basis, n-pentane decreases viscosity more effectively than HCFC-142b.

4.2.2 Inert Gases

Among the alternative blowing agents tested, inert gases (Table 4.4) have received considerable attention due to their low cost and mild impact on the environment. As with any other blowing agents, inert gases will readily plasticize polymer melts. Globally, viscosity is a function of both polymer and gas type. Despite great advantages, processing foam with inert gases does have a few technical drawbacks. Generally, these small molecules are harder to process than conventional blowing agents, owing to their low solubility and high diffusivity.

Figure 4.2 depicts the solubility of known blowing agents in PS at 373 K, measured at equilibrium conditions [19]. It is clear that the solubility of carbon dioxide is much lower than all the other conventional gases so that higher pressures need to be applied in order to dissolve an equivalent number of molecules. This feature might cause processing limitations for foam

TABLE 4.4

Physical Properties of Inert Gases

Name	Formula	GWP	Vapor Pressure (kPa) at 25°C	Boiling Point (°C)	Vapor Thermal Conductivity at 25°C (mW/mK)
Air	N₂, O₂	—	—	−195	26
Nitrogen	N₂	—	—	−195	26.1
Nitrous oxide	N₂O	310	5729	−89	17
Carbon dioxide	CO₂	1	6448	−78.5	16.6
Carbon monoxide	CO	—	—	−192	24
Argon	Ar	—	—	−186	17.2
Oxygen	O₂	—	—	−183	26

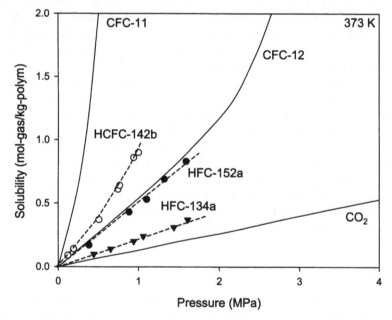

FIGURE 4.2

Comparison of solubilities of various gases in polystyrene at 373 K. (From Sato, Y. et al., *Polym. Eng. Sci.*, 40, 1369, 2000 [19]. With permission.)

extrusion where pressure both in the extruder and in the shaping die is limited. Since minimal reachable density is directly correlated with gas concentration (Equation 4.3), lower densities can be more easily achieved with less volatile agents. Additionally, carbon dioxide diffuses rapidly out of the polymer matrix, which may cause problems in controlling the foam growth process, resulting in surface defects such as corrugation.

In the extrusion process, equilibrium solubility derived from sorption measurements does not solely govern the amount of available gas. The dynamic

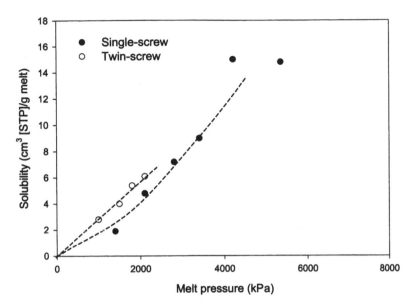

FIGURE 4.3
Comparison of solubility of CO_2 in PS at 215°C obtained from single-screw (1.32 kg/hr) and twin-screw (2.4 kg/hr) experiments. (From Zhang, Q. et al., *J. Cell. Plast.*, 37, 537, 2001 [20]. With permission.)

nature of extrusion, which is a function of mixing efficiency and residence time, may actually limit or enhance the actual gas concentration in the polymer melt [20]. Figure 4.3 depicts gas solubility data obtained from a single-screw or twin-screw extruder. As reported by Zhang et al. [20], it appears that the mixing efficiency is superior in a twin-screw extruder and that a higher concentration of gas may be incorporated even though the polymer throughput is actually higher for the twin screw. The data for the twin-screw extruder depicted in Figure 4.3 gives a slope equivalent to the Henry's law constant for equilibrium. This indicates that inadequate homogeneity of the gas-laden melt in the single screw is probably responsible for the poor consistency with the equilibrium measurements. In the twin-screw extruder, the screws were configured with specific elements for improved mixing. This would contribute significantly to enhanced gas dissolution and improved homogeneity.

Residence time is also a key factor for allowing dissolution of the blowing agent, as seen by the effect of polymer throughput on gas solubility (Figure 4.4). At low screw speed (30 rpm), when the polymer feed rate is high, the degree of fill is high and the residence time is reduced; as a result, the mixing efficiency is relatively poor. Moreover, increasing polymer throughput would directly increase the shear rate and might help to trigger bubble nucleation. So the correlation between gas solubility measured in Figure 4.4 might be an indication of the effect of shear. But shear also causes heating of the polymer melt. Michaeli and Heinz [21] basically came to the

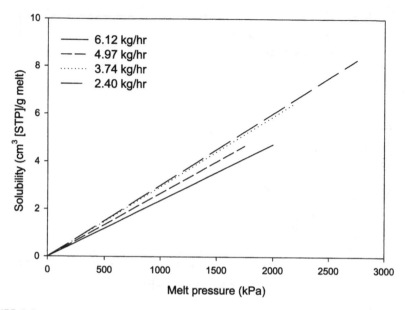

FIGURE 4.4
Effects of polymer throughput on solubility of CO_2 in PS at 215°C (twin-screw extruder at 30 rpm). (From Zhang, Q. et al., *J. Cell. Plast.*, 37, 537, 2001 [20]. With permission.)

same conclusion when they studied the effect of processing on the solubility of CO_2 in thermoplastic polyurethane (TPU) as their results showed that the solubility limit decreases with increasing screw speed for all screw configurations. Not only does the higher output reduce the contact time between polymer and blowing agent, it also causes higher shear stress and heating that increases melt temperature and further contributes to lower solubility.

Generating high or medium density foams can be easily accomplished with inert gases [22,23]. However, producing low-density foams is far more challenging. Several attempts have been made to improve the processability of carbon dioxide for foaming thermoplastic polymers. Most strategies have been aimed at either increasing the solubility or reducing the diffusivity of the gas. Strategies to increase gas solubility in polymer melts include increasing free volume through the addition of a plasticizer, adding bulky constituents on the polymer chains, and generating favorable interactions with the gas (CO_2-philic interactions). In that latter case, specific interactions may also help to reduce diffusion. Wilkes et al. [24] claim that adding polyglycol or polyglycol ether to a polyethylenic extrudate can help to control foaming, due to their gas absorbing properties, which slow the diffusion of carbon dioxide. Vo and Paquet [25] claim that adding an oxygen-containing monomer to polystyrene enables operation pressures to be significantly reduced (higher gas solubility) when using CO_2 as the blowing agent. The researchers state that 40% of oxygen-containing additive may be added to polystyrene. As additives may have a measurable impact on gas solubility, it appears that the major problem lies in the amount that is required to obtain a true improvement.

Carbonyl-containing additives were tested in polystyrene to improve processing with CO_2 [26]. According to Kazarian et al. [27], carbonyl groups interact specifically with the gas so that they can increase solubility and reduce diffusivity simultaneously. Results obtained by Vachon and Tatibouët [26] showed that such additives have a small but measurable effect on gas solubility. Similarly, these additives were found to reduce gas diffusion at room temperature. It should be emphasized that the amount of additive required to obtain a significant improvement reached up to 10 wt%. In such cases, the foamed material may have characteristics (rheological or mechanical) far different from those desired.

One successful avenue for the use of inert gases has been the production of microcellular foams. The concept for microcellular plastics was created in response to the industrial need for reducing material cost for certain polymer products without major compromise to mechanical properties [28]. The main idea lies in creating bubbles with dimensions smaller than pre-existing flaws. It is claimed that such materials would have improved mechanical properties from the original unfoamed polymer, along with a significant density reduction. By definition, microcellular foams have cell sizes in the order of 10 μm or below and a cell density greater than 10^9 cells per cubic centimeter of unfoamed polymer. To achieve such small cell sizes and high densities, a sudden thermodynamic instability — either a temperature increase, a pressure drop, or a combination of the two — is required to induce bubble nucleation from the gas–polymer solution.

In a batch process, microcellular foams are produced by charging a polymer in a pressure vessel with the physical blowing agent, usually nitrogen or carbon dioxide, at constant pressure until the sorption equilibrium is reached. The polymer can be expanded then either by rapid decompression of a heated pressure vessel or by submitting the solid saturated sample to a higher temperature environment such as an oil bath followed by subsequent quenching. The latter approach was used to produce microcellular foams made from copolymers of ethylene and octene, as shown in Figure 4.5 [29]. Continuous production of microcellular foams using extrusion has also been successfully attempted [30,31]. Park et al. [31] were able to produce microcellular foams made of high-impact polystyrene (HIPS) and a CO_2 concentration of 5 wt% with cell densities of 10^{10} cells/cm^3 and controlled expansion ratios. In order to generate high cell nucleated densities, a special pressure-drop nucleation nozzle was developed. Globally lowering the melt and nozzle temperature helped to control coalescence, and cell sizes in the 10 μm range were reported. However, reduced cell sizes were obtained at the expense of density reduction.

The MuCell® technology (Trexel, Inc., Woburn, MA) uses modified extrusion and injection molding processes [32,33]. With microcellular foam technology, the polymer melt is obtained by introducing a supercritical fluid (SCF) through the injectors. In the mixing section of the screw, the gas is

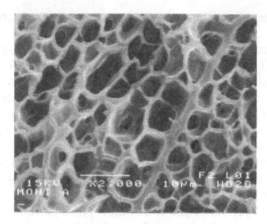

FIGURE 4.5
Micrograph of microcellular foam sample obtained from ethylene and octene copolymer (24 wt% octene) foamed with CO_2 at 90°C for 30 s.

quickly dispersed and mixed with the polymer to form a single-phase solution. In extrusion, the foaming process can be obtained with either a tandem line or a single extruder (twin- or single-screw). The next critical step in the continuous production of microcellular plastics is promotion of high bubble nucleation rates in the polymer–gas solution by means of a rapid pressure drop. The greatest possible number of cells for a given pressure difference would to be nucleated out of a given polymer–gas solution if the pressure drops instantaneously because of the greater thermodynamic instability [28]. However, in reality, the pressure drops over a finite period of time.

Zotefoams, Inc., (Croydon, U.K.) has developed an industrial batch process where nitrogen pressures up to 69 MPa are used to produce low-density polyolefin foams [34]. This technology uses a microcellular pre-expansion step followed by another expansion stage. First, a polymer sheet is extruded and crosslinked by peroxides or irradiation. Second, the slabs are pre-expanded in an autoclave to produce a microcellular foam having an initial density of 720 kg/m³ and cell diameters smaller than 10 μm. In the final step, the slabs containing gas-filled nuclei are further expanded by heating at atmospheric pressure.

4.2.3 Hydrofluorocarbons

HFCs are among the compounds that have been identified for the replacement of HCFCs in thermal insulation applications. Two major uses for HFCs are polyurethane rigid foam and thermoplastic foam extrusion. HCFC-141b was used as a preferred blowing agent for polyurethane (PU) foams while HCFC-142b was used for polystyrene foam board manufacturing. Both of these blowing agents will be banned in the next few years and replacements are currently being developed for specific market applications. Table 4.5 lists

TABLE 4.5

Physical Properties of Hydrofluorocarbons (HFCs)

Name	Formula	GWP$_{100\,yr}$	Vapor Pressure (kPa) at 25°C	Boiling Point (°C)	Vapor Flame Limits (% vol)	Vapor Thermal Conductivity at 25°C (mW/mK)	Ref.
HFC-23	CHF$_3$	12,000	4742	−82.1	None	14	35
HFC-32	CH$_2$F$_2$	550	1686	−53.1	13.6–28.4	12.0	36
HFC-41	CH$_3$F	97	3802	−78.4	Yes	—	—
HFC-125	C$_2$HF$_5$	3,400	1267	−48.5	None	13.4	36
HFC-134	CHF$_2$CHF$_2$	1,100	522.6	−19.7	None	—	37
HFC-134a	CH$_2$FCF$_3$	1,300	597.2	−26.1	None	12.0	38–40
HFC-143	CH$_2$FCHF$_2$	330	876	5	Probable yes	10.1	41
HFC-143a	CF$_3$CH$_3$	4,300	1272	−47.8	6.0–n/a	12.1	39
HFC-152	CH$_2$FCH$_2$F	43	82.1	30.7	Slightly	—	—
HFC-152a	CH$_3$CHF$_2$	120	601.1	−24.7	3.9–16.9	13.6	—
HFC-227ea	CF$_3$CHFCF$_3$	3,500	450	−17	None	13.3	40, 42
HFC-236ca	CHF$_2$CF$_2$CHF$_2$	—	—	12.6	None	—	40, 42
HFC-236cb	CH$_2$FCF$_2$CF$_3$	1,300	—	−1.4	None	—	40, 42
HFC-236ea	CHF$_2$CHFCF$_3$	1,200	208.1	6.5	None	14.2	40, 42–44
HFC-236fa	CF$_3$CH$_2$CF$_3$	9,400	272.3	−1.1	None	12.5	40, 42
HFC-245ca	CH$_2$FCF$_2$CHF$_2$	640	100.7	25	None	13.5 or 13.2	40, 42, 44–46
HFC-245cb	CF$_3$CF$_2$CH$_3$	—	—	−18.3	None	13.1	40, 42, 44, 46
HFC-245fa	CHF$_2$CH$_2$CF$_3$	950	152	15.3	None	12.2	40, 42, 44, 46
HFC-356mfc	CF$_3$CF$_2$CH$_2$CH$_2$F	—	—	24.6	—	—	—
HFC-356mffm	CF$_3$CH$_2$CH$_2$CF$_3$	—	—		None	11.8	46, 47
HFC-365mfc	CF$_3$CH$_2$CF$_2$CH$_3$	890	80	40.1	3.5–9.0	10.6	—
HFC-43-10mee	CF$_3$CFHCFHCF$_2$CF$_3$	1,300	31.3	55	None	—	48

Source: GWP data taken from the Intergovernmental Panel on Climate Change, *Third Assessment Report: Climate Change 2001* [9].

common HFCs. Although the actual number of HFCs is far more extensive than the few compounds listed in Table 4.5, the number of HFCs presently developed and used as foam blowing agents is actually reduced because of technical and commercial limitations. In the last few years, two candidates have emerged as possible replacements for HCFC-141b in polyurethane foams, namely HFC-365mfc and HFC-245fa.

The choice of HFC-365mfc and HFC-245fa lies in the manufacturing possibilities and the availability of raw materials, pentachlorobutane and pentachloropropane. It is the accessibility of the compounds and the costs of production that make the use of these blowing agents viable [49].

In the polystyrene foam board market, HCFC-142b will also have to be replaced by non-ozone-depleting substances, such as HFC-134a and HFC-152a. However, other suitable candidates may include HFC-245fa [50]. Although that particular blowing agent was developed for implementation in the PU industry, its physical properties, which are intermediate to HCFC-142b and n-pentane, render it suitable for thermoplastic foam extrusion.

Another common hydrofluorocarbon, HFC-134a, was also processed in polystyrene as well as low-density polyethylene. Unlike HFC-245fa, which requires the use of nucleating agents to obtain suitable cell structure and density, HFC-134a has been reported to self-nucleate easily at moderate concentration [51]. Thus, low-density PS foams with a fine cell structure were produced without using a nucleating agent. Figure 4.6 shows micrographs of foam samples made from similar concentrations of HFC-134a or HFC-245fa without any nucleating agent. The average cell diameter is around 30 μm for foams nucleated with HFC-134a and 300 μm for HFC-245fa. Although the molecular weights of the blowing agents are slightly different (102 g/mole for HFC-134a vs. 134 g/mole for HFC-245fa) and the actual amount of gas molecules differs from one formulation to the other, the nucleation density is much more important for foam samples produced with HFC-134a using similar processing conditions.

In low-density polyethylene, larger cells (with lower nucleation density) are generally obtained for a given BA. Figure 4.7 depicts the cell size distribution of LDPE foams produced with various concentrations of HFC-134a. As expected, increasing the amount of BA in the polymer clearly decreases cell size. However, the decrease in cell size also has a detrimental effect on the number of open cells and on final density. Once nucleate density reaches a critical number, the polymer walls become too thin to sustain the deformation, which leads to an important fraction of open cells and partial collapse of the foam. This is due to differences between the stabilization processes of amorphous vs. semicrystalline polymers. In amorphous polymers, the stiffness of the cell structures is caused by the sudden viscosity increase of the polymer matrix following gas depletion. In PE, as well as other polyolefins, viscosity is not greatly affected by the presence of the BA. It is rather the melt strength of the polymer matrix that sustains cell deformation. Obviously, as the number of cells increase, the amount of material in the cell wall decreases, which in the end, limits the stress that the matrix

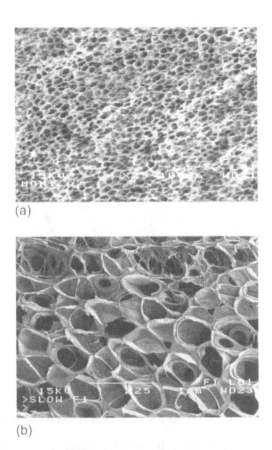

FIGURE 4.6
Micrographs of polystyrene foam samples blown with (a) 7.1 wt% HFC-134a ($\rho = 55$ kg/m^3) or (b) 7 wt% HFC-245fa ($\rho = 59$ kg/m^3) at a melt temperature of 150°C.

can sustain. Thus, low-density foams made from polyolefins are generally obtained at the expense of cell size reduction.

4.2.4 Other PFAs

Water is very appealing because it is so abundant and cheap. From many perspectives (economical, environmental, toxicological), it would make the ideal BA if it could meet the proper requirements for foaming. There are very limited scientific data on water, and those that are available would indicate that it has a very limited potential. Water was used as a blowing agent for foaming thermoplastic elastomers (TPEs) [52,53]. Typically, foams produced had a minimal density above 100 kg/m^3 owing to the poor solubility of water in TPE. According to Sahnoune [53], increasing water content above 2.5 wt% does not lower foam density further. In fact, the melt becomes saturated with water and the BA dispersion is not fully completed. This in

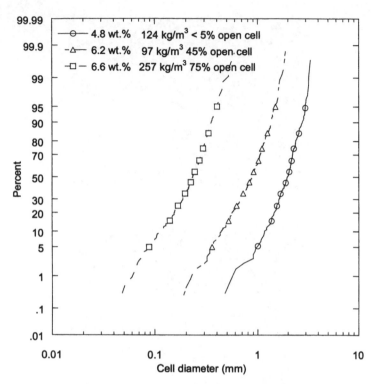

FIGURE 4.7
Cell size distribution of LDPE foams made with various HFC-134a concentrations (wt%).

turn leads to larger cells and to cell rupture. Pallay and Berghmans [54] used water to make expandable polystyrene beads. Because water is not soluble enough in PS, starch was added to act as a carrier for the blowing agent. Unfortunately, the morphology of the foams produced was irregular because of the incompatibility of the two phases. Improvement was made possible by grafting maleic anhydride on the polymer backbone.

Like water, volatile liquids, such as ketones, alcohols, and ethers, can be used as BAs for thermoplastic foams, although they are often used in combination with other BAs. For instance, low molecular weight alcohols have boiling points that are simply too high to be used as single blowing agents. Extruding PS with isopropanol generates exclusively high-density foams [55]. In the case of 2-ethyl hexanol (T_b = 182°C) [56], it was not possible to obtain any kind of foam when using the pure BA. In both cases, a volatile co-agent was necessary in order to reach acceptable densities.

Innovative blowing agents are now being developed for the PU industry. Their physical properties might also render them suitable for thermoplastic foam extrusion. Methylal (CH_3-O-CH_2-O-CH_3) has been very recently proposed as a suitable alternative to HCFC-141b. It has a boiling point of 42.3°C but can be readily mixed with HFC-134a, hydrocarbons, CO_2, or even HFC-245fa to customize blowing power [57]. Formic acid was also proposed

TABLE 4.6

Physical Properties of Hydrofluoroethers

Name	Formula	Vapor Pressure (kPa) at 25°C	Boiling Point (°C)	Flash Point (°C)	Vapor Thermal Conductivity (mW/mK) at 50°C
HFE-245mf	$CF_3CH_2OCH_2F$	86.9	29	None	13.75
HFE-347mfc	$CF_3CF_2CH_2OCHF_2$	44.9	45.94	None	12.93
HFE-347pc-f	$CHF_2CF_2OCH_2CF_3$	28.4	56.22	None	12.37

Source: Data from Takada, N. et al., *J. Cell. Plast.*, 35, 389, 1999 [59].

TABLE 4.7

Boiling Points of Fluoroiodocarbons

Name	Formula	Molecular Weight	Boiling Point (°C)
Trifluoroiodomethane	CF_3I	195.9	−22.5
Pentafluoroiodoethane	CF_3CF_2I	245.9	12
Difluoroiodomethane	CHF_2I	177.9	21.6
1,1,1,2,3,3,3-Heptafluoro-2-iodopropane	CF_3CFICF_3	295.9	39
1,1,2,2-Tetrafluoro-1-iodoethane	CF_2ICHF_2	227.9	40
1,1,2,2,3,3,3-Heptafluoro-1-iodopropane	$CF_3CF_2CF_2I$	295.9	41
Pentafluoroiodopropane	$C_3H_2F_5I$	187.4	52
Fluoroiodomethane	CH_2FI	159.9	53.4

Source: Data from Nimitz, J., *Proc. Polyurethanes 1994 Conf.*, 1994, 110 [60].

as a co-blowing agent for rigid polyurethane foams [58]. Other classes of compounds that might find potential use in thermoplastic foams include hydrofluoroethers (HFE) and fluoroiodocarbons (FICs). HFEs and FICs have rather high boiling points [59,60]. Implementation in the thermoplastic foam industry would thus require blending. Tables 4.6 and 4.7 report typical physical properties of HFEs and FICs.

4.2.5 Blends of Physical Blowing Agents

Mixtures of blowing agents are being used to replace conventional gases. Since single agents may not perform adequately, it may be suitable to combine two or more agents to match the desired properties. These properties are numerous and vary as a function of the final product. Blending can efficiently reduce cost, customize blowing power, improve solubility, and increase product performance (i.e., dimensional stability, thermal insulation, etc.).

In PU, blends developed have been based mostly on HFC-245fa or HFC-365mfc. Bogdan and Williams blended HFC-245fa with water or hydrocarbons [61,62]. HFC-365mfc has been successfully blended with HFC-134a, HFC-245fa, HFC-227ea, and pentane isomers [49,63,64]. Although all of these blowing agents might be used for thermoplastic foaming, these blends were developed with boiling points that were suitable for PU foaming only. Different blend compositions made from the same compounds might find practical use for thermoplastic foam extrusion.

In the PS foam board industry, blending was first introduced to duplicate the physical properties of CFCs. For instance, blends of HCFC-142b/HCFC-22 or HCFC-142b/ethyl chloride were used in replacement of CFC-12 [65,66]. Despite their good performance as blowing agents, HCFCs will soon have to be replaced with zero ODP substitutes such as HFCs. Unfortunately, it appears that HFC-134a for instance has a lower solubility in PS than do common HCFCs, and therefore may not be easily used as the sole blowing agent in the production of low-density extruded polystyrene boards [66]. In that particular case, adding a more soluble co-agent can help to reach targeted densities. As an example, Albouy et al. [67] have used blends of HFC-134a and cyclopentane to produce insulating foam boards. The advantages are multiple, as the use of cyclopentane should not only reduce cost but improve global solubility in PS, thus enabling lower foam densities compared to those obtained with HFC-134a alone [68]. In another work, global foam morphology was improved by blending HFC-134a with a minor phase of isopropanol [55]. It was reported that low-density PS foams were obtained without the presence of large blow holes as seen in foam samples produced with only HFC-134a. The addition of a minor phase of isopropanol was shown to improve the homogeneity of the polymer–gas solution.

Similarly, inert gases such as carbon dioxide have a low solubility in polymers and require equipment modifications to accommodate their high vapor pressures. In most cases, inert gases benefit from the addition of a co-BA. When CO_2, for instance, is combined with a low-volatility gas, the more volatile component drives the degassing pressure. Therefore, adding a co-agent can globally increase the total concentration without increasing the degassing pressure, which is known to limit the process. Daigneault et al. [56] combined 2-ethyl hexanol with CO_2 to produce low-density PS foams. In fact, by using an optimal concentration of both, they were able to produce foam samples with densities as low as 29 kg/m³, whereas foams produced with only CO_2 were limited to 34 kg/m³. It is clear from the few examples mentioned above that blending blowing agents is one very interesting avenue from many technological standpoints. However, the effect of blending on basic foam processing is not very well understood. This will more thoroughly investigated in Section 4.5.

4.3   Chemical Blowing Agents

4.3.1   General Properties

Chemical blowing agents (CBAs) are made of a class of solid or liquid compounds that decompose to form gases under processing conditions. Most CBAs are solids and decompose within a given temperature range. According to Shutov [69], there are various requirements that have to be met when choosing a CBA:

- Decomposition temperature of the CBA should be slightly higher than the melt or processing temperature of the material.
- Gas must be liberated within a narrow temperature range.
- Rate of gas liberation must be sufficient and controllable.
- Gas and decomposition products must be noncorrosive, nontoxic, and nonflammable.
- CBA must be adequately dispersed in the polymer matrix.
- Polymer matrix should not be destroyed by CBA or heat generated during decomposition.
- Internal gas pressure should be controlled.
- Diffusion rate of the gases generated by CBA should be adequate for the chosen polymer matrix.

One advantage of using CBAs is that they do not require any modification of the existing equipment. Thus, they can be simply compounded and added with the polymer matrix in the hopper. Fine-tuning of the temperature profile may be required in order to prevent premature decomposition of the CBA. In extrusion, it is important that thermal decomposition occurs after the melt seal on the screw so that no gas is lost through the hopper. In injection molding, the polymer melt containing the CBA is melted on the screw at a temperature low enough to prevent decomposition. Once molten, the polymer is extruded in a mold maintained at a higher temperature where the CBA is decomposed into gas [70].

The available grades of CBAs are based on different chemical compounds and have different gas yields (the amount of gas released by the decomposition of the blowing agent) and different gas profiles. For instance, chemicals that release more carbon dioxide will produce lower density foams, since carbon dioxide is more soluble in thermoplastics than nitrogen. The solubility of a gas is governed by Henry's law (Equation 1.1) and the Henry law's constant is 1.5 to 4 times greater for carbon dioxide than nitrogen in common polymers. Therefore, it is necessary to apply more pressure in order to keep nitrogen dissolved in the molten polymer. This may cause some limiting processing conditions and density reduction.

Two main classes of CBAs can be found: endothermic and exothermic agents. Endothermic agents absorb energy during decomposition, which imparts some kind of auto-cooling to the process. In some cases, it may be necessary to generate heat in order to ensure complete reaction. The most common endothermic blowing agent is a blend of citric acid and sodium bicarbonate. It reacts to form water vapor, carbon dioxide, and a solid residue (sodium citrate dehydrate). The reaction takes place at two temperature ranges, 160 and 210°C, and the gas yield is 120 cc/gm at STP.

Exothermic agents, on the other hand, release heat and allow a more rapid and efficient decomposition of the chemical agent. However, external cooling may be required to stabilize the foam. Common commercial exothermic blowing agents are listed in Table 4.8. The proper choice of a CBA depends on the decomposition temperature and the gas yield of the compound.

4.3.2   Foaming with CBAs

Producing low-density foams with CBAs is fairly straightforward provided that the gas yield and solubility/diffusivity properties of the CBA are chosen with relation to the processed polymer. However, other additives are generally used in combination with these agents to optimize processing conditions. For instance, activators are combined with CBAs to trigger decomposition at the desired temperature range. Another important parameter that needs to be controlled is the rheology of the polymer resin. It is clear from the work of Gendron and Vachon [72] that the viscosity of the resin and its resistance to deformation will determine the lowest reachable density. Hence, crosslinking agents can be used to modify resin properties and help to decrease density further.

CBAs can be combined to improve performance. In fact, Luebke [73] proposes that there are significant advantages in combining endothermic and exothermic blowing agents. This could lead to a more controlled release of heat that might otherwise be detrimental to the properties of the polymer matrix. CBAs are also used in combination with other BAs to modify global foam morphology. Because these substances are solids that form gas nuclei and a non-negligible fraction of solid residue, they also act as nucleating agents. Champagne et al. [74] investigated the effect of azodicarbonamide on nucleation and density reduction in polyolefin foams blown with CO_2. Figure 4.8 depicts the foam densities obtained in polyolefin blends foamed with or without added azodicarbonamide (AZ).

It can be observed that, in the presence of AZ, the lowest achievable density is obtained at a lower CO_2 concentration. Of course, AZ releases additional gas, but these clusters also impact the foam nucleation. According to the authors, the effective concentration of AZ was kept at 0.1 wt%. Considering that only a third of the weight will be converted into gas, it is clear that the AZ does not contribute to an increase in the blowing power. However, the presence of these gas clusters and solid particles can efficiently trigger the

TABLE 4.8

Processing Conditions for Common Exothermic Blowing Agents [10,71]

Grade	Chemical Compound	Gas Composition		Solid Decomposition Product	Gas Yield (cc gas at STP)	Decomposition Temperature Range (°C)
AZ	Azodicarbonamide	N_2 CO CO_2 NH_3	65% 24% 5% 5%	Urazol Biurea Cyamelide Cyanuric Acid	220	205–215
RA	p-Toluene sulfonyl semicarbazide	N_2 CO_2 NH_3 CO	55% 37% 3% 2%	Ditolyl disulfide Ammonium p-toluene sulfonate	140	228–235
OT	p,p′-Oxybis (benzenesulfonyl hydrazide)	N_2 H_2O	91% 9%	Nonpolar aromatic sulfur-containing polymer	125	158–160
TSH	p-Toluene sulfonyl hydrazide	N_2 H_2O	n/a n/a	Ditolyl disulfide p-Tolyl-p-toluene thiosulfinate p-Toluenesulfinic acid hydrazine p-Toluenesulfonyl hydrazide salt of p-toluene-sulfonic acid	115	110–120

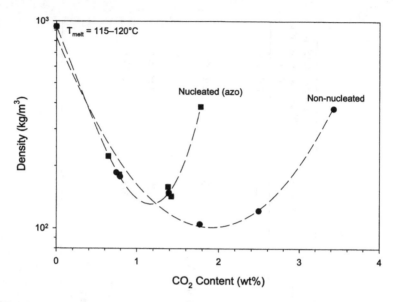

FIGURE 4.8
Density of 40/60 LDPE–HDPE blends foamed using neat CO_2. Melt temperature was set at 115 to 120°C. Lines are only intended to be used as guides. (From Champagne, M.F. et al., *Proc. ANTEC Society of Plastics Engineers*, 3125, 2004 [74]. With permission.)

nucleation process and foam growth. Therefore, smaller cells are obtained with AZ for a constant concentration of CO_2. As the gas concentration is increased further, the results suggest that there comes a point where the cell walls become too thin to sustain deformation, leading to foam densification.

4.4 Pathways for PFA Validation

There are many techniques that can be used to assess the processability of blowing agents. Readers may refer to other chapters of this book for in-depth reviews of solubility measurements, ultrasonic monitoring, and extensional rheology. These topics will not be covered in detail in this chapter but rather exploited to probe the behavior of single and binary blowing agent systems in later sections.

4.4.1 Monitoring of Gas–Polymer Systems

In order to determine the suitability of a BA for foam extrusion, all stages of the foaming process must be considered. The first step includes incorporation of the BA in the polymer, and care should be taken to assure complete dissolution and homogenization. Once the blowing agent is dissolved, general polymer–gas parameters such as melt viscosity and pressure need to be evaluated in order to control processing and prevent gas loss. In the final step, the melt is extruded from the shaping die, releasing the BA and inducing nucleation. Die geometry, pressure drop, and pressure drop rate all have a critical impact on initial nucleation, followed by foam growth, which is affected by the nature of the BA and its concentration in the melt.

Solubility is a parameter that is most often evaluated in gas–polymer systems as it will limit the technological capabilities of a particular BA. This parameter can be determined by many experimental methods, including gravimetric techniques, vapor pressure methods, spectroscopic or ultrasonic monitoring, and rheological measurements [75]. Gravimetric measurements are straightforward and can be conveniently carried out by using a high-precision electrobalance (described in Chapter 1) capable of measurements over a wide range of temperatures and pressures. In that case, the polymer–gas system is allowed to reach equilibrium conditions and the microbalance directly measures the weight of the gas dissolved in the polymer matrix. The sensitivity of the apparatus can reach 1 part per million, which makes this technique ideal for the solubility measurement of gases with very low solubility (i.e., inert gases). This technique is also suitable for measuring the solubility of gases in polymers either in the rubbery or glassy states. Diffusivity measurements can also be derived from the gas uptake vs. time curves. Because of the sensitivity of these balances, corrections must be made for changes in buoyancy of the sample. Thus, knowledge of the dilation of the polymer with BA uptake is also needed.

A variation on this approach uses a closed pressure vessel for measurements of gas solubility. Known masses of polymer and gas are introduced into the vessel. The temperature and the pressure of the BA can be varied. Solubility measurement is made by weighing the pressure vessel and its contents. Using the following equation, along with knowledge of the densities of each component inside the reactor, allows for measurement of solubility [75]:

$$V_t = \frac{w_p}{\rho_p} + \frac{w_g - w_{l,g}}{\rho_{g,g}} + \frac{w_{l,g}}{\rho_{l,g}} \tag{4.4}$$

where V_t is the volume of the pressure vessel, w_p is the mass of the polymer, w_g is the mass of the gas, $w_{l,g}$ is the mass of the dissolved gas, ρ_p is the density of the polymer, $\rho_{g,g}$ is the density of the gas phase, and $\rho_{l,g}$ is the density of the dissolved gas.

In the absence of dilatometric data, the density of the dissolved gas can be assumed to be similar to the density of the polymer [76]. An important advantage of this technique is that solubility can be measured at very high temperatures and pressures depending on the rating of the vessel.

Spectroscopic techniques such as in-line infrared (IR) monitoring can be directly performed in the processing stream to determine the actual amount

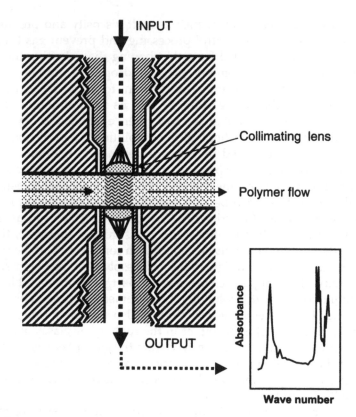

FIGURE 4.9
Schematic of instrumented die used for in-line infrared monitoring.

of gas dissolved in the polymer melt. Figure 4.9 illustrates an instrumented die that can be equipped with dual transmission IR sensors and placed at the end of a foam extrusion line. A gear pump follows the slit die to control pressure fluctuations. The dimensions of the slit can also be varied to accommodate the desired stress and pressure levels. These types of sensors used for in-line monitoring are linked with fiber-optic cables to Fourier-transform infrared (FTIR) spectrophotometer, which records spectra of the melt in the near-infrared (NIR) region. The advantage of using IR sensors is that they may be used to track several components individually in the polymer melt. Additionally, IR sensors may be used to probe the phase separation of the gas from the polymer melt during the foam nucleation process [77].

Ultrasonic sensors (USs) monitor sound propagation through the polymer melt. These sensors can be used both in-line and off-line to provide information on the phase behavior of the polymer melt such as composition and homogeneity. Off-line measurements can even be used at a variety of temperature and pressures ranges that exceed the pressure and temperature ranges of an extrusion line. Hence, data can be obtained at cryogenic conditions up to polymer degradation. As detailed in Chapter 5, information

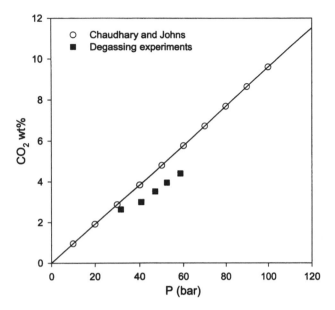

FIGURE 4.10
Solubility of CO_2 in LDPE at 150°C as determined from degassing pressures and equilibrium solubility data. (Data from Chaudhary, B.I. and Johns, A.I., *J. Cell. Plast.*, 34, 312, 1998 [76].)

on specific volume changes can also be monitored and can give access to the direct measurement of the glass transition temperature.

One of the main uses of in-line ultrasonic monitoring is the study of the degassing pressure for foam processing. The degassing pressure is the pressure at which the gas comes out of the polymer matrix during a controlled pressure drop. It has been found to be highly dependent on experimental parameters such as temperature, BA type and concentration, presence of nucleating agent, and the rheological behavior of the polymer matrix. In systems comprising only the BA and the polymer, degassing pressures were found to be very informative on the solubility behavior of the gas. In agreement with static solubility measurements, for a fixed temperature, the degassing pressure increases with concentration. A comparison between recorded degassing pressures and static measurements is given in Figure 4.10 for various concentrations of CO_2 in LDPE.

On that basis, the gas solubility appears to be lower in the extrusion process, which may be caused by many factors. First, the dynamic nature of extrusion may not allow the dissolution of the maximum amount of gas in the time frame of the experiment. On average, the residence time in an extruder will range in terms of minutes while off-line static experiments may require hours or days to reach equilibrium (see Figure 1.8). Errors may also arise from temperature differences between the actual melt temperature and the temperature read by the thermocouples. Finally, one last important parameter is the effectiveness of the screw configuration on the dissolution process [78].

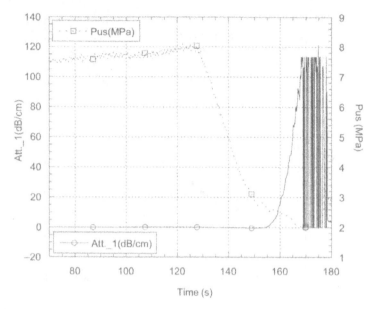

FIGURE 4.11

Attenuation and pressure signals recorded for 6 wt% HFC-245fa in polystyrene at 190°C. (From Vachon, C. and Gendron, R., *Cell. Polym.*, 22, 295, 2003 [50]. With permission.)

Solubility can be evaluated through the measurement of degassing pressures. As mentioned previously, these pressures may not represent the true equilibrium solubility of a gas but may actually represent limiting solubility conditions met in the processing line. Mixing efficiency, temperature profiles, screw configuration, and residence time all affect the amount of gas that can be dissolved. Additional data can be extracted from the ultrasonic attenuation in order to provide information on the dispersion of the BA in the polymer melt. A stable signal prior to degassing is a clear indication of a fully homogeneous system (from 0 to 130 s in Figure 4.11). As the pressure is being gradually decreased in the die (starting from 130 s), there comes a point where the system can no longer maintain the gas dissolved, and phase separation occurs. This is evidenced by the sharp increase in the attenuation signal caused by the scattering effect of the gas bubbles. The scattering becomes more and more important as the bubbles grow in size. Because sound attenuation is highly sensitive to the presence of various phases, noise present in the signal even before complete phase separation can be indicative of heterogeneous conditions (Figure 4.12).

4.4.2 Foam Characterization

An interesting tool that has been recently exploited is the analysis of foam extrusion profiles. These profiles are the result of a successive line of events during the foaming process. A typical illustration of a foam sample produced

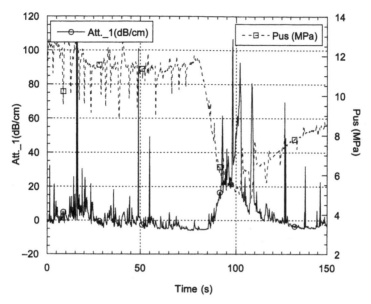

FIGURE 4.12
Attenuation and pressure signals recorded for 12 wt% HFC-245fa in PS at 190°C. (From Vachon, C. and Gendron, R., *Cell. Polym.*, 22, 295, 2003 [50]. With permission.)

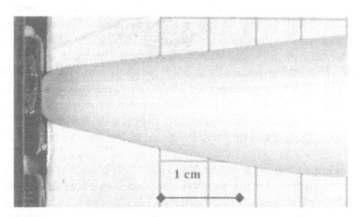

FIGURE 4.13
Photograph of polystyrene foam expanded with 5 wt% HFC-245fa and 0.5 wt% talc at 150°C.

by extrusion is given in Figure 4.13. First, a swell is observed right at the exit of the shaping die due to the elastic recovery of the molten polymer. In the second stage, nucleation is triggered rapidly followed by foam expansion, which is greatly dependent on BA concentration and volatility. PS foams expanded with CO_2, for instance, will grow rapidly right after the exit of the shaping die, while foams made with pentane will expand more slowly. But other additives contained in the formulation may play a determining

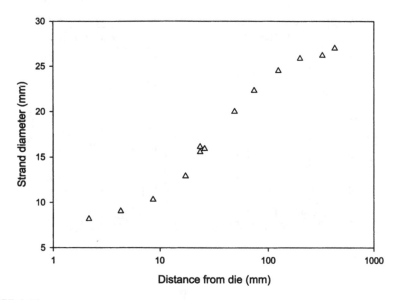

FIGURE 4.14
Expansion profile of polystyrene foam expanding with 5 wt% HFC-245fa and 0.5 wt% talc at 150°C.

role. For instance, nucleating agents impact the expansion by triggering the nucleation process sooner. In some cases, this may lead to a more efficient use of the dissolved BA and to lower foam densities. The final stage of the foam extrusion profile illustrated in Figure 4.14 (starting at a distance of 100 mm) is indicative of the stabilization step where the foam is cooled into its final shape. Processing parameters such as melt and die temperature are especially important as evidence of shrinkage and collapse can be determined by looking at the profiles. Globally, foam profile analysis may be of some use for optimizing the extrusion process or finding suitable foam formulations. Examples of this will be given in the following sections of this chapter.

Final validation can be accomplished through standard characterization techniques, including density analysis, open-cell fraction, and morphology. Determination of density can be obtained by geometrical calculations or experimental methods (geometric pycnometry, water displacement, etc.). Open-cell fraction is determined using a pycnometer where a gas (generally nitrogen) is injected in a calibrated chamber containing a known volume of foam. The difference between the theoretical available volume and the volume of gas actually injected in the chamber corresponds to the small fraction of gas that has penetrated the open cells. Hence, the open-cell fraction can be determined through basic calculations. Thorough analysis of the foam structure remains more complex. Basic micrographs of the structure can be obtained through scanning electronic microscopy, but the analytical treatment of these images remains challenging. Although computer image analysis software is currently available, the main difficulty resides in clearly

TABLE 4.9

Standard Test Methods for Cellular Materials

Test Method Designation	Test Method Description	Ref.
Apparent density	Density is obtained by measuring the weight and volume of a sample.	79
Open-cell content	Method determines numerical values for open cells. Porosity is assessed by measuring the accessible volume of a material.	80, 81
Cell size	Determination of the apparent cell size and rigid cellular plastics by counting the number of cell-wall intersections in a specified distance.	82
Water absorption	Method covers the determination of the water absorption of rigid cellular plastics by measuring the change in buoyant force resulting from immersion.	83

discriminating the cells from the struts. One way to counter this problem is to draw the cell contours by hand and analyze the resulting image with the software. Equivalent cell diameter, cell volume, strut thickness, and even cell shape factors can be calculated and used for foam characterization. Common standard test methods used for cellular materials are provided by the American Society for Testing and Materials under the designations D1622, D2842, D2856, D3576, and D6226 (Table 4.9).

4.5 Blowing Agent Processing

4.5.1 Plasticization

Blending BAs affects foam processing in many ways. Viscosity, for instance, will be impacted differently from one type of BA to another. Ultrasonic monitoring performed on PS melts plasticized with either carbon dioxide or HFC-134a clearly shows that on an equivalent molar basis, HFC-134a has a greater plasticizing effect than CO_2 (Figure 4.15). Simultaneous independent results obtained from conventional rheological measurements are in agreement with what was observed by the ultrasonic sensors. At a constant shear rate, shear stress gradually decreases as a function of gas concentration with lower values obtained for PS/HFC-134a. An increase of 50% in molar concentration of the carbon dioxide would be required to match the viscosity decrease induced by a given amount of HFC-134a (Figure 4.16). Globally, larger BA molecules tend to plasticize polymer melts more efficiently than small ones, probably owing to a more important spacing of the polymer chains. This plasticization can be observed by the depreciation of the glass transition temperature (T_g) for doped systems. Figure 4.17 reports variations of T_g on an equivalent molar basis for different gases. It can be clearly seen

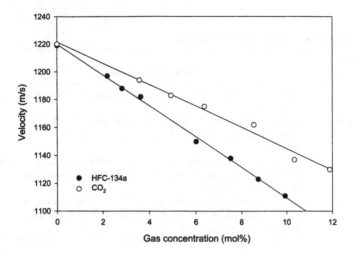

FIGURE 4.15
Sound velocity (homogeneous phase) of polystyrene–blowing agent mixtures as a function of concentration. Measurements were conducted at a melt temperature of 190°C. All data normalized at P = 8.3 MPa. (From Vachon, C. and Gendron, R., *Cell. Polym.*, 22, 75, 2003 [51]. With permission.)

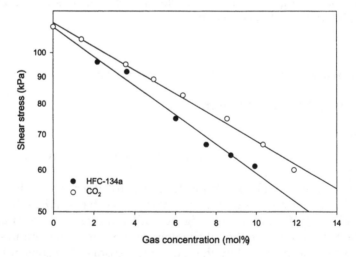

FIGURE 4.16
Semi-logarithmic plot of shear stress as a function of blowing agent concentration. Measurements were conducted at a melt temperature of 190°C. (From Vachon, C. and Gendron, R., *Cell. Polym.*, 22, 75, 2003 [51]. With permission.)

that the lowering of T_g is directly correlated to molecular weight, i.e., molecular size.

Polymer type will also greatly influence the extent of plasticization. In comparison, plasticization by HFC-245fa was evaluated both in LDPE and in PS. Figure 4.18 reports the sound velocity and shear stress of

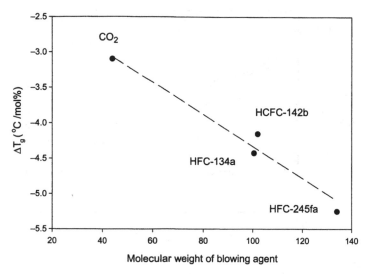

FIGURE 4.17
Glass transition lowering of polystyrene per molar percentage of gas as a function of molecular weight. (From Vachon, C. and Gendron, R., *Cell. Polym.*, 22, 295, 2003 [50]. With permission.)

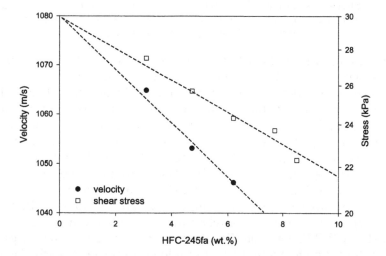

FIGURE 4.18
Shear stress and sound velocity of mixtures of HFC-245fa in low-density polyethylene. Measurements were done at a melt temperature of 190°C. Sound velocity was corrected for a pressure of 8.3 MPa. (From Vachon, C., *Cell. Polym.*, 23, 109, 2004 [84]. With permission.)

LDPE/HFC-245fa mixtures at 190°C prior to degassing. It can be observed that both sound velocity and stress decrease with concentration indicating that HFC-245fa plasticizes the polymer matrix. Results previously reported for the same blowing agent in PS [50] showed that the decrease of sound velocity and shear stress was not as important in LDPE as in PS. Sound

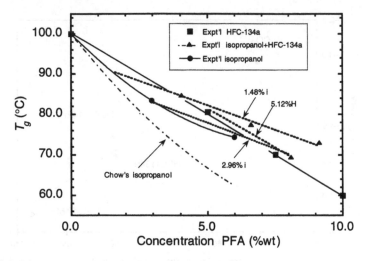

FIGURE 4.19

Variation of the glass transition temperature (T_g) as a function of various mixture concentrations of blowing agents isopropanol and HFC-134a. The thin slash-dotted line corresponds to Chow's estimate for isopropanol. The solid line corresponds to experimental results for mixtures of polystyrene (PS) with isopropanol and PS with HFC-134a. The bold dotted line corresponds to mixtures of PS with isopropanol and HFC-134a, where one of the two diluents is maintained at a constant fraction — for example, 5.12%H stands for 5.12 wt% of HFC-134a (H stands for HFC-134a, and i for isopropanol) combined with a variable concentration of isopropanol. (From Gendron, R. et al., *Cell. Polym.*, 23, 1, 2004 [55]. With permission.)

velocity change associated with the plasticizing effect of HFC-245fa in PS is 7.6 m/s/wt%, while it is only of 6.03 m/s/wt% in LDPE.

Plasticization studies were also performed on the binary systems of blowing agents. Of course, blending can offer some technological or economical advantage, but the main interest is to understand how these agents interact with one another, and probe these systems for possible synergistic effects. An example of that is given in the rheological study of blends based on HFC-134a and isopropanol in polystyrene. The analysis of the raw data was performed according to a procedure detailed in Gendron et al. [85] and includes corrections for pressure, temperature, and the Rabinowitsch factor for the apparent shear rate. Modeling of the results through the Williams-Landel-Ferry (WLF) equation (Equation 2.7) finally leads to the corresponding glass transition temperature that should be associated to each plasticized system, according to its PFA composition (Figure 4.19).

For the polymer systems containing only isopropanol, the T_g computed using Chow's predictive equation based on the molecular weight of the solvent and of the repeat unit of the polymer [85,86] are indicated using a thin slash-dotted line. The solid and dotted lines correspond, respectively, to mixtures of PS with isopropanol only, and to different systems with one diluent at a fixed concentration. The agreement between Chow's estimates and experimental results for HFC-134a is excellent over the investigated range of concentrations. Experimental results for mixtures of PS with isopropanol exhibit T_g values much higher than those predicted by Chow's equation. According to results derived from Chow's equation and based solely on its molecular weight, isopropanol (M_w = 60.096 g/mol) should be a better plasticizer than HFC-134a (M_w = 102.3 g/mol). Through experimental results, we see that in fact, both HFC-134a and isopropanol behave similarly.

Several combinations of isopropanol and HFC-134a have been attempted, and many observations can be made on the rheological response of these mixtures. The first two sets of data under investigation are those containing a fixed fraction of isopropanol (1.48 and 2.96 wt%), while the concentration of HFC-134a is varied. These are illustrated in Figure 4.19 by the two dotted lines labeled *1.48% i* and *2.96% i*. The two lines are quite parallel, and their slopes are smaller than that of sole HFC-134a. The T_g for mixtures involving the two diluents are always higher than the results obtained for HFC-134a, at identical total concentrations. Similar conclusions can be drawn for the set of data involving a fixed fraction of isopropanol. This example shows that combinations of BAs do not always behave in ways predictable from the properties of the pure components. In this particular system, the combination of the two diluents seem to present some interactions that slightly inhibit the plasticizing effect.

Studies of the plasticization of gas–polymer systems probed by ultrasonic and rheological measurements are generally in agreement: lowering of shear stress is accompanied by lowering of ultrasonic velocity, as was seen for single BA systems (CO_2 or HFC-134a in PS). However, an unusual discrepancy has been recently noted in blends of CO_2 and (cyclohexane or isopropanol) in PS. Figure 4.20(a) and (b) shows the variation of shear stress and sound velocity as a function of CO_2 ratio in equivalent molar blends.

Figure 4.20(a) and (b) shows that when cyclohexane or isopropanol is gradually replaced by CO_2, shear stress and ultrasonic velocity vary in opposite ways. It would be expected from previous studies that larger molecules are better plasticizers and should cause a more important drop of viscosity. In that sense, the macroscopic data provided by the on-line slit die follow that trend. The ultrasonic results, which are more sensitive to the microscopic entanglement of the polymer chain, suggest something entirely different. As described in Chapter 5, sound velocity V_{us} is directly proportional to the elastic modulus (L') and specific volume (V_{sp}) (or density [ρ]) of the polymer–gas solution, following Equation 4.5:

$$V_{us} = \sqrt{L'/\rho} = \sqrt{L' \cdot V_{sp}} \tag{4.5}$$

In accordance with the rheological measurements reported in Figure 4.20(a), which measure the pressure drop (i.e., viscous loss) encountered in the slit die, it is expected that the elastic modulus will also be lower for mixtures containing increasing proportions of cyclohexane or isopropanol. However, when measuring the sound velocity, one has to account for the

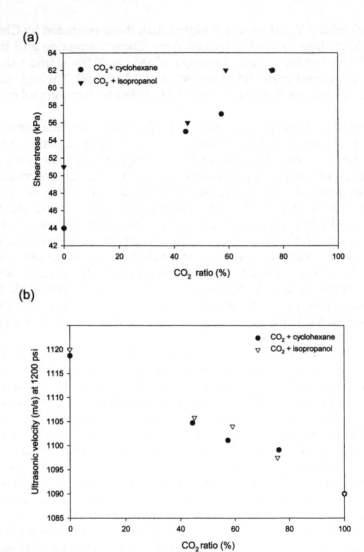

FIGURE 4.20
Shear stress (a) and sound velocity (b) of equivalent molar blends of CO_2 and isopropanol or CO_2 and cyclohexane. Ultrasonic velocity was corrected at 8.3 MPa.

effect of specific volume. It may be postulated that even at constant pressure, the swelling of the polymer–gas solution is so important in the presence of cyclohexane or isopropanol that, overall, the sound velocity is higher than with pure CO_2.

4.5.2   Solubility

Solubility has been investigated for single and dual BA systems through the measurement of degassing pressures. Figure 4.21 shows that the degassing

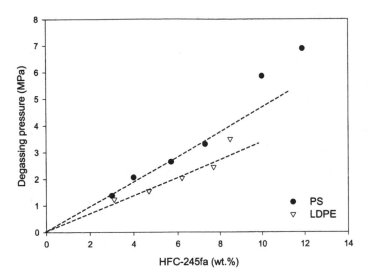

FIGURE 4.21
Degassing pressures recorded at 190°C for HFC-245fa in LDPE and PS. (From Yachon, C., *Cell Polym.*, 23, 109, 2004 [84]. With permission.)

pressures of HFC-245fa are systematically lower in LDPE than in PS, in agreement with known equilibrium solubility measurements for other polymer–gas systems [87,88]. The data in PS also show that higher pressures than expected were recorded at 10 and 12 wt% HFC-245fa and were interpreted as an indication of heterogeneous conditions, also evidenced by the noise in the attenuation signal (Figure 4.12).

In the case of HFC-245fa in LDPE, no evidence of heterogeneity was seen either in the degassing pressures or the original sound attenuation signals, although HFC-245fa concentrations investigated in LDPE were kept below 10 wt%. The results reported for HFC-245fa show that the apparent solubility coefficients measured at 190°C and 1 MPa are 0.159 mol gas/kg polymer in PS and 0.229 mol gas/kg polymer in LDPE. These results are in the same range as those obtained for HCFC-22 by Gorski et al. under similar pressure and temperature conditions: 0.17 mol gas/kg polymer in PS, 0.25 mol gas/kg polymer in HDPE, and 0.30 mol gas/kg polymer in PP [88].

Solubility behavior was also investigated for binary systems. Figure 4.22 depicts the degassing pressures recorded for pure HFC-134a and blends of HFC-134a and isopropanol in PS at various temperatures. Degassing pressures increase with concentration in agreement with classical solubility measurements (see Chapter 1). When isopropanol is added to HFC-134a, the parabolic curve is shifted to lower temperatures and slightly lower pressures. Adding a second BA should increase degassing pressures or at least maintain similar degassing pressures when a low volatility agent is used, such as in the case of isopropanol. Therefore, small differences seen here may fall within experimental error. The horizontal shift however is a clear indication that

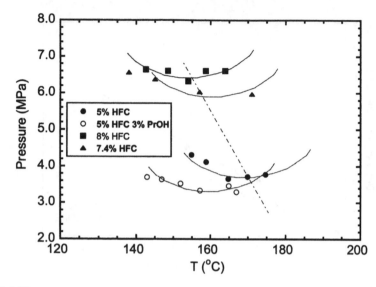

FIGURE 4.22
Degassing pressures as a function of the melt temperature for pure HFC-134a and HFC-134a blended with isopropanol. (From Gendron, R. et al., *Cell. Polym.*, 23, 1, 2004 [55]. With permission.)

isopropanol acts as a plasticizer in the blend and allows mixtures to be processed at lower temperatures. Similar conclusions were drawn for the CO_2–ethanol system. In both cases, carbon dioxide was blended with a low-volatility agent where one component acts essentially as a plasticizer while the other also generates foam growth.

In the case of HFC-134a and CO_2, both gases can be considered highly volatile. Blends of HFC-134a and CO_2 in PS yield degassing pressures intermediate to the pure gases (Figure 4.23). A closer look at Figure 4.23 shows that higher degassing pressures are noted for pure CO_2, in the range of 6–7 MPa. The degassing pressures for neat HFC-134a are roughly 1 MPa below those of carbon dioxide. Generally, increasing the relative proportion of HFC-134a in the blend decreases the corresponding degassing pressures. Degassing results for the blends containing 3.02 mol% CO_2 + 5.30 mol% 134a diverge from the overall trend, with unusually high degassing pressures. This set of experiments has been performed with a higher PFA molar fraction, i.e., 8.3 mol% vs. 7.3 mol% for the other trials. This approximately 10% concentration increase may explain the proportional rise of the measured degassing pressures. One interesting point of comparison between this system and the previous one is that there is no important shift of the parabola along the temperature axis when adding a co-agent. The major shift is rather observed along the pressure range (y axis), indicating that both gases contribute to foam growth.

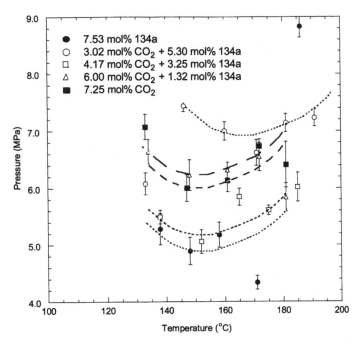

FIGURE 4.23
Degassing pressure as a function of temperature for equivalent molar blends. A curve for each data set has been plotted as a guide for the eye only. (From Vachon, C. and Gendron, R., *Cell. Polym.*, 22, 75, 2003 [51]. With permission.)

4.5.3 Nucleation and Expansion

Expansion profiles have recently shown to be very informative on foam processing. Whereas classical morphology measurements give access to the final structure, foam profiles can yield pertinent information on initial nucleation and growth. Figure 4.24 shows the foam expansion profiles of HFC-245fa in PS obtained for various gas concentrations and talc loadings. It can be seen that both talc and BA concentrations significantly affect foam growth. At the lowest concentration tested (5 wt% HFC-245fa) and in absence of talc, foam expansion is extremely limited and high densities are thus expected. Minimal amounts of talc (0.5 wt%) must be added to allow suitable nucleation and bubble growth. Increasing the talc concentration even more (1 wt%) accelerates the expansion process further but does not impact final foam diameter.

A similar behavior is observed by increasing the amount of BA, which, as with adding talc, provides additional nucleation sites. This is particularly obvious for foam expansion profiles without talc as results show that an optimal combination of both is required to obtain maximum expansion. For example, at 5 wt% HFC-245fa, expansion profiles are markedly different for

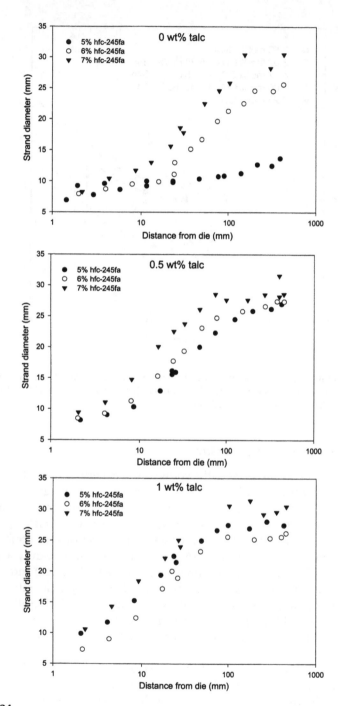

FIGURE 4.24

Expansion profiles of foam samples made from polystyrene and HFC-245fa. (From Vachon, C. and Gendron, R., *Cell. Polym.*, 22, 295, 2003 [50]. With permission.)

various talc loadings while at higher blowing agent concentration (6 wt% and 7 wt%), expansion profiles are similar for talc loadings of 0.5 wt% and 1 wt%. Adding a nucleating agent is generally regarded as a means of providing adequate cell diameter and structure. However, from looking at the expansion profiles, nucleating agents do more than just impact morphology: they clearly modify the entire bubble growth kinetics, which permits a more efficient use of the dissolved gas.

In the case of PS, where viscosity is greatly affected by temperature and the amount of BA (i.e., plasticization), cell growth and stabilization is generally assured by the viscosity increase caused by the escaping BA and the cooling of the polymer melt. Increasing the amount of nucleation sites allows foam growth to occur sooner in the same time frame. Therefore the presence of talc particles will accelerate the initial nucleation process at a time where the mixture has a lower viscosity (i.e., a higher temperature). This phenomenon may explain in part a more efficient expansion of the PS. Once the initial nucleation is launched, bubble growth will be highly dependent on the blowing power of the gas molecules, which may phase separate or remain dissolved for a longer period in the polymer melt. Providing more nucleation sites favors a quicker release of the remaining BA, also improving foam expansion. In accordance with foam expansion profiles, lower densities were noted for larger strand diameters [50].

Expansion profiles in LDPE differ from those reported in PS in many ways. First, foam growth is clearly accelerated with increasing BA concentration in PS [50], while it remains slow in LDPE. As shown in Figure 4.21, degassing pressures are lower in LDPE, and therefore, a higher concentration of HFC-245fa is required to generate an equivalent propensity to nucleate. As an example, a concentration of 8 wt% HFC-245fa in LDPE is required to match the degassing pressure of 5 wt% gas in PS. Second, the impact of talc on the foam growth kinetics is more important in LDPE. Figure 4.25 compares the expansion profiles of LDPE and PS foams extruded with a similar gas concentration and temperature. Whereas 0.25 wt% talc is sufficient to induce the maximum growth rate in LDPE, 1 wt% talc is required to generate a similar effect in PS, although both sets of trials were conducted with exactly the same nucleating agent.

Binary blends were also investigated using foam profile analysis. Results are given for HFC-134a, CO_2, and mixtures of the two gases in PS (Figure 4.26). Despite small differences, both BAs and their blends have very similar expansion profiles. In all cases, the expansion is abrupt and maximum expansion is obtained around 10 mm. This behavior is significantly different than that observed with HFC-245fa in PS, owing to its much slower diffusion rate. Larger diameters of exiting foams are obtained for foams extruded with only CO_2. As the proportion of HFC-134a increases in the blend, diameters become smaller. In the case of the foam samples extruded with only HFC-134a, there is even some evidence of shrinkage as foam diameter tends to decrease slightly over distance. It should be emphasized that although foam samples were extruded at the same temperature, the plasticizing effect of

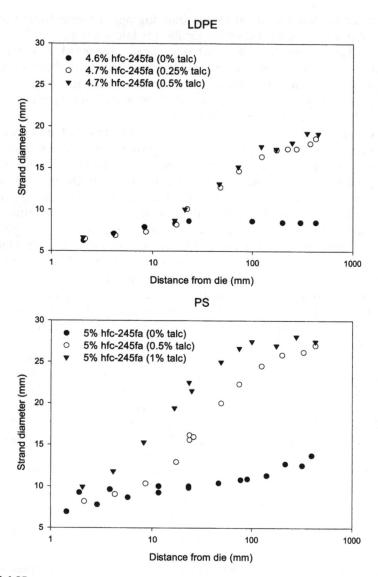

FIGURE 4.25
Foam extrusion profiles of LDPE and PS extruded with similar concentrations of HFC-245fa. Melt temperature was 150°C in LDPE and ranged between 148–155°C in PS. (From Vachon, C., *Cell Polym.*, 23, 109, 2004 [84]. With permission.)

each blowing agent is slightly different. As previously demonstrated in Figures 4.15 and 4.16, HFC-134a has a greater plasticizing effect than CO_2, and therefore a "softer" foam may lead to partial collapse and cell wall rupture. Moreover, since wall thickness is inversely proportional to the nucleation density, a thinner wall will be prone to rupture, which should result in the formation of more open cells.

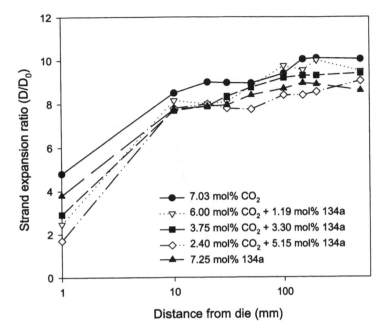

FIGURE 4.26
Strand expansion ratio for foam samples extruded at constant temperature, $T = 149°C$. (From Vachon, C. and Gendron, R., *Cell. Polym.*, 22, 75, 2003 [51]. With permission.)

Formulations based on equivalent molar blends of BA generated foams with similar densities, except for the foam samples made of 100% HFC-134a, where shrinkage is likely responsible for the slight density increase [51]. Micrographs of the foam samples were obtained and analyzed in order to extract cell size distributions (Figure 4.27). A constant decrease in the cell sizes is seen with increasing proportion of HFC-134a in the blend. The global morphology of the foams is characterized by a narrow cell size distribution as seen by the steepness of the slope. A slightly wider distribution is noted for the blend containing nearly equal proportions of CO_2 and HFC-134a. The general trend seen in the cell sizes and even degassing pressures would suggest that gas molecules evolve as a whole and do not form independent clusters exclusively comprised of one type of gas.

For a constant amount of gas molecules dissolved in the polymer melt (7.2 mol%), a higher number of small bubbles are produced with HFC-134a. This feature is also reflected in the measurement of β, the nucleate cell density, which represents the number of nucleation sites per cm^3 of unfoamed polymer [70]:

$$\beta = \frac{6\left(\rho_{PS}/\rho_{foam} - 1\right)}{\pi d_{cell}^3} \qquad (4.6)$$

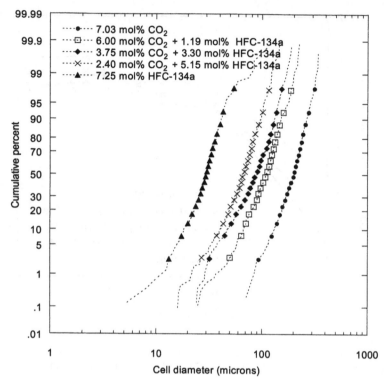

FIGURE 4.27
Cell size distribution of foam samples extruded at constant temperature. (From Vachon, C. and Gendron, R., *Cell. Polym.*, 22, 75, 2003 [51]. With permission.)

where ρ is the density of the polymer and foam, and d_{cell} is the average cell diameter.

When comparing pure CO_2 and HFC-134a, β is almost 1000 times more important for HFC-134a. In other terms, for a fixed amount of BA, and in absence of significant cell coalescence, the quantity of gas molecules is 1000 times less important in each cell for HFC-134a. The nucleate cell density (β) was plotted as a function of the relative proportion of each BA in the blend (close symbols, Figure 4.28).

Micrographs of the foams are also shown in the same figure. Correction of the nucleate cell density for those blends that deviate from the nominal concentration of 7.2 mol% total, especially the mixture made of 2.40 mol% CO_2 and 5.15 mol% HFC-134a, has also been attempted. The correction performed, based on the assumption that nucleation density is an exponential function of the PFA concentration, is represented by the open symbols in Figure 4.28. The large difference in the nucleation densities reported in Figure 4.28 between the foams obtained with the pure PFAs can be explained by the differences in their diffusion coefficients.

Once nucleation is launched, one must consider the competition that exists between subsequent nucleation from still-dissolved PFA molecules and the

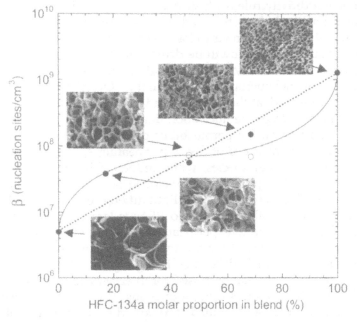

FIGURE 4.28
Nucleate cell density (β) as a function of relative proportion of HFC-134a in the blend. Experimental data points are represented by closed symbols, while open symbols correspond to results corrected on a basis of equally set molar concentrations (details can be found in the text). Micrographs of the resulting foams made from equivalent molar concentrations of blowing agent are also included (same magnification for all). (From Vachon, C. and Gendron, R., *Cell. Polym.*, 22, 75, 2003 [51]. With permission.)

cell growth of existing cells. The latter is closely related to the diffusion of the gas molecules into existing adjacent nucleation sites [89], while the former originates from the favorable conditions finally met in case of lower PFA molecule concentrations. Huge differences between the diffusion coefficients of HFC-134a and those of carbon dioxide in PS have been reported [90], with those of HFC-134a being 1000 times smaller, even if the PS matrix experiences more plasticization with HFC-134a. In that case, the extra free volume generated through dissolution of HFC-134a does not necessarily increase the mobility of the solvent molecules, as indicated by the large difference in their diffusion coefficients. When considering the formulations having a constant amount of gas molecules, the higher diffusion of CO_2 would likely contribute to a fast migration of PFA molecules toward existing nucleation sites, and a rapid expansion of the resulting foam. Because of its higher solubility, HFC-134a has a lower propensity to phase-separate. And because of its smaller diffusion coefficients, HFC-134a molecules are less prone to "feed" existing bubbles within the time frame of the gradual pressure reduction, and the still-dissolved PFA would favor additional nucleation taking place, under milder conditions.

A simple log-additivity rule is illustrated by a dotted line in Figure 4.28. Another relationship is hypothesized and is represented by a reverse S-shaped curve that was drawn very close to the experimental and corrected data points. Should bubble growth be driven by the slow diffusion of gas molecules, these results would indicate that small amounts of HFC-134a greatly affect the overall nucleation of CO_2. In the CO_2-rich blend, nucleation is significantly enhanced; at the opposite end, a small amount of CO_2 in an HFC-134a-rich blend drops the nucleation density drastically. Ettouney and Majeed [91], reporting on the permeability of mixtures of gases in membranes, claim that the presence of faster permeating species enhances the permeation rates of slower species, while reduction in the faster species' permeability is caused by presence of the slower permeating species. In the present case, even though there exists a slight difference in the solubility, the difference in the permeability of the two PFAs is largely dictated by their diffusion coefficient. Respective permeability coefficients of both carbon dioxide and HFC-134a, when they are mixed, may tend to come close to average numbers, with their difference being significantly reduced. This could explain the quasi-plateau proposed for intermediate concentrations in Figure 4.28. This observation is quite interesting since it would mean that overall nucleation could be controlled through adequate mixtures of gases having different permeability coefficients. In addition, since interaction between the two PFAs is assumed to affect diffusivity and be responsible for the behavior illustrated in Figure 4.28, it could also be hypothesized that solubility, as represented by the degassing curves of Figure 4.23, is sensitive to the interactions between the HFC-134a and CO_2 molecules, which leads to deviations such as those displayed in Figure 4.28.

Another example of unexpected foam morphology is reported in the work of Daigneault and Gendron [56], where CO_2 was blended with 2-ethyl hexanol (2-EH) in PS. At 8.6 wt% CO_2 and 1.6 wt% 2-EH, an intriguing bimodal cell size distribution was reported, with larger cells having an average diameter of 250 μm and smaller cells having a diameter of 25 μm (Figure 4.29). Because 2-EH has a high boiling point (182°C), low processing temperatures may lead to phase separation from the matrix. It was proposed that the partitioning coefficient favors a higher concentration of CO_2 in the 2-EH phase than in the matrix. This higher dissolution of the CO_2 in the 2-EH phase would be a major contributing factor to the importance of foam void fraction ascribed to the larger bubbles. Nucleation takes place as the expandable PS exits the die. If there is a 2-EH phase, CO_2 bubble growth in that 2-EH phase will be rapid, and coalescence will take place. The larger bubbles would therefore originate from the 2-EH phase while the smaller bubbles would originate from the CO_2 phase.

Gendron and Daigneault also reported bimodal cell size distributions in polycarbonate foams [30]. In that particular case, the final morphology was clearly due to the processing conditions and the low pressure in the shaping die. Hence, using a similar gas concentration and melt temperature, they were able to eliminate the bimodal distribution by increasing the pressure

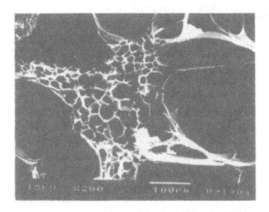

FIGURE 4.29
Micrograph of polystyrene foam extruded with a blend of 8.6 wt% CO_2 and 1.6 wt% 2-EH. (From Daigneault, L.E. and Gendron, R., *J. Cell. Plast.*, 37, 262, 2001 [56]. With permission.)

at the die, which inhibited premature degassing. In other cases, bimodal cell size distributions were created on purpose. Arora et al. [92] used a sequential depressurization profile to induce bimodal distribution in microcellular polystyrene foams. It may be expected that adding more decompression steps may produce foams with three or more cell size distributions.

4.6 Conclusions

Numerous BAs are available for thermoplastic foam extrusion. An overview of the various agents used and their physical properties was presented. Based on the examples summarized in this chapter, it is clear that these properties affect how each agent should be handled and processed. Therefore, this requires adequate monitoring of the gas–polymer melt and the subsequent foaming step. Moreover, it was shown that combinations of BAs sometimes lead to an intriguing behavior that should be thoroughly investigated.

But the debate over finding non-ozone–depleting substances for foaming needs to be further extended. As the first generation of alternative agents was developed following the adoption of the Montreal Protocol, the upcoming Kyoto Protocol, which is in the process of being adopted worldwide, will pose another problem: that of reducing global warming. Therefore, fluorinated derivatives and hydrocarbons currently used for foam extrusion may eventually be scrutinized or even banned. In light of this, it is expected that more effort will be aimed at using atmospheric gases for extrusion foaming. The development of novel and innovative foam structures (vacuum-insulated panels, layered foams, etc.) is another option that should be pursued, especially for applications requiring insulating properties.

Abbreviations

AZ:	Azodicarbonamide
BA:	Blowing agent
CBA:	Chemical blowing agent
CFCs:	Chlorofluorocarbons
2-EH:	2-Ethyl hexanol
FICs:	Fluoroiodocarbons
FTIR:	Fourier-transform infrared
GWP:	Global warming potential
HCs:	Hydrocarbons
HCFCs:	Hydrochlorofluorocarbons
HFCs:	Hydrofluorocarbons
HFEs:	Hydrofluoroethers
HIPS:	High-impact polystyrene
IR:	Infrared
LDPE:	Low-density polyethylene
NIR:	Near-infrared
ODP:	Ozone-depleting potential
PE:	Polyethylene
PFA:	Physical foaming agent
PS:	Polystyrene
PU:	Polyurethane
SCF:	Supercritical fluid
TPE:	Thermoplastic elastomer
TPU:	Thermoplastic polyurethane
US:	Ultrasonic sensor
UV:	Ultraviolet
WLF equation:	Williams–Landel–Ferry equation

References

1. Bird, G., Ed., United Nations Environment Programme (UNEP), in Protecting the ozone layer, Volume 4, Foams, 2001.
2. Midgley, T., Henne, A.L., and McNary, R.R., U.S. Patent 1,833,847, 1931.
3. Midgley, T., Henne, A.L., and McNary, R.R., U.S. Patent 1,930,129, 1933.

4. Skochdopole, R.E., **Cellular materials**, in *Encyclopedia of Polymer Science and Technology: Plastics, Resins, Rubbers, Fibers*, Vol. 3, Mark, H.F., Gaylord, N.G., and Bikales, N.M., Eds., Interscience Publishers, 1964, 80–130.
5. Munters, C.G. and Tandberg, J.G., U.S. Patent 2,023,204, 1935.
6. Johnson, F.L., U.S. Patent 2,256,483, 1941.
7. Molina, M.J. and Rowland, F.S., Stratospheric sink for chlorofluoromethanes: chlorine atom-catalysed destruction of ozone, *Nature*, 249, 810, 1974.
8. United Nations Environment Programme (UNEP), Ozone Secretariat, *Montreal Protocol on Substances that Deplete the Ozone Layer*, 2000.
9. Intergovernmental Panel on Climate Change, *Climate Change 2001: The Scientific Basis*, Cambridge University Press, Cambridge, U.K., 2001, chap. 6.
10. Throne, J.L., Foaming agents, in *Thermoplastic Foams*, Sherwood Technologies, New York, 1996, chap. 4.
11. Norton, P., Chemical, physical, and thermal properties of gases, in *The CRC Handbook of Thermal Engineering*, Kreith, F., Ed., CRC Press, Boca Raton, 2000, Appendix A.
12. ANSI/ASHRAE Standard 24-2001: Designation and safety classification of refrigerants, American Society of Heating, Refrigerating and Air-Conditioning Engineers, Inc., Atlanta.
13. Pontiff, T., Foaming agents for foam extrusion, in *Foam Extrusion*, Lee, S.T., Ed., Technomic, Lancaster, PA, 2000, chap. 10.
14. Behravesh, A.H., Park, C.B., and Venter, R.D., Challenge to the production of low-density, fine-cell HDPE foams using CO_2, *Cell. Polym.*, 17, 309, 1998.
15. Park, C.B., Liu, Y., and Nagui, H.E., Challenge to fortyfold expansion of biodegradable polyester foams by using carbon dioxide as a blowing agent, *J. Vinyl Addit. Technol.*, 6, 39, 2000.
16. Xanthos, M. et al., Parameters affecting extrusion foaming of PET by gas injection, *J. Cell. Plast.*, 36, 102, 2000.
17. Park, C.P., Stabilisation of polyolefin foams during ageing, in *Proc. Blowing Agents Foaming Processes 2003*, Rapra Technology Ltd., Shawbury, 2003, 223.
18. Gendron, R. and Champagne, M.F., Effect of physical foaming agents on the viscosity of various polyolefin resins, *J. Cell. Plast.*, 40, 131, 2004.
19. Sato, Y. et al., Solubility of hydrofluorocarbon (HFC-134a, HFC-152a) and hydrochlorofluorocarbon (HCFC-142b) blowing agents in polystyrene, *Polym. Eng. Sci.*, 40, 1369, 2000.
20. Zhang, Q., Xanthos, M., and Key, S.K., Dissolution of atmospheric gases in polystyrene melt in single and twin-screw foaming extruders, *J. Cell. Plast.*, 37, 537, 2001.
21. Michaeli, W. and Heinz, R., Foam extrusion of thermoplastic polyurethanes (TPU) using CO_2 as a blowing agent, *Macromol. Mater. Eng.*, 284, 35, 2000.
22. Dey, S.K. et al., Use of inert gases in extruded medium density polypropylene foams, *J. Vinyl Technol.*, 2, 339, 1996.
23. Dey, S.K., Jacob, C., and Xanthos, M., Inert-gas extrusion of rigid PVC foam, *J. Vinyl Technol.*, 2, 48, 1996.
24. Wilkes, G.R et al., U.S. Patent 5,905,098, 1999.
25. Vo, C.V. and Paquet, A.N., U.S. Patent 5,426,125, 1995.
26. Vachon, C. and Tatibouët, J., Effect of additives on the solubility and diffusivity of CO_2 in polystyrene, in *Proc. ANTEC 2004*, Society of Plastics Engineers, Chicago, 2004, 2527.

27. Kazarian, S.G. et al., Specific intermolecular interaction of carbon dioxide with polymers, *J. Am. Chem. Soc.*, 118, 1729, 1996.
28. Park, C.B., Continuous production of high-density and low-density microcellular plastics in extrusion, in *Foam Extrusion*, Lee, S.T., Ed., Technomic, Lancaster, PA, 2000, chap. 11.
29. Vachon, C. and Gendron, R., Effect of gamma-irradiation on the foaming behavior of ethylene-co-octene polymers, *Radiat. Phys. Chem.*, 66, 415, 2003.
30. Gendron, R. and Daigneault, L.E., Continuous extrusion of microcellular polycarbonate, *Polym. Eng. Sci.*, 43, 1361, 2003.
31. Park, C.B., Behravesh, A.H., and Venter, R.D., Low density microcellular foam processing in extrusion using CO_2, *Polym. Eng. Sci.*, 38, 1812, 1998.
32. Vanvuchelen, J. et al., Microcellular PVC foam for thin wall profile, *J. Cell. Plast.*, 36, 148, 2000.
33. Xu, J. and Kishvaugh, L., Simple modeling of the mechanical properties with part weight reduction for microcellular foam plastic, *J. Cell. Plast.*, 39, 29, 2003.
34. Eaves, D.E. and Witten, N., Product and process developments in the nitrogen autoclave process for polyolefin foam manufacture, in *Proc. ANTEC 1998*, Society of Plastics Engineers, Atlanta, 1998, 1843.
35. Shan, Z., Penoncello, S.G., and Jacobsen, R. T., A generalized model for viscosity and thermal conductivity of trifluoromethane (R-23), *ASHRAE Trans.: Symp.*, 36, 757, 2000.
36. Tanaka, Y., Matsuo, S., and Taya, S., Gaseous thermal conductivity of difluoromethane (HFC-32), pentafluoroethane (HFC-125), and their mixtures, *Int. J. Thermophys.*, 16, 121, 1995.
37. Tamatsu, R., Sato, H., and Watanabe, K., Measurements of pressure-volume-temperature properties of 1,1,2,2-tetrafluoroethane, *J. Chem. Eng. Data*, 37, 216, 1992.
38. Krueger, D.C. and Reichel, C.J., 1,1,1,2-Tetrafluoroethane as the primary blowing agent for rigid polyurethane foams using conventional mixing equipment, *J. Cell. Plast.*, 30, 164, 1994.
39. Tanaka, Y., Nakata, M., and Makita, T., Thermal conductivity of gaseous HFC-134a, HFC-143a, HCFC-141b, and HCFC-142b, *Int. J. Thermophys.*, 12, 949, 1991.
40. Göktun, S., An overview of chlorine-free refrigerants for centrifugal chillers, *Energy*, 20, 937, 1995.
41. Perkins, R. et al., Thermal conductivities of alternatives to CFC-11 for foam insulation, *J. Chem. Eng. Data*, 46, 428, 2001.
42. Beyerlein, A.L., DesMarteau, D.D., and Hwang, S.H., Physical properties of fluorinated propane and butane derivatives as alternative refrigerants, *ASHRAE Trans.*, 99, 368, 1993.
43. Di Nicola, G. and Giuliani, G., Vapor pressure and P-V-T measurements for 1,1,1,2,3,3-hexafluoropropane (R-236ea), *J. Chem. Eng. Data*, 45, 1075, 2000.
44. Dohrn, R., Treckmann, R., and Heinemann, T., Vapor-phase thermal conductivity of 1,1,1,2,2-pentafluoropropane, 1,1,1,3,3-pentafluoropropane, 1,1,2,2,3-pentafluoropropane and carbon dioxide, *Fluid Phase Equil.*, 158-160, 1021, 1999.
45. Sharpe, J. et al., Evaluation of HFC 245ca and HFC 236ea as foam blowing agents, *J. Cell. Plast.*, 31, 313, 1995.
46. Heinemann, T. et al., Experimental determination of the vapor phase thermal conductivity of blowing agents for polyurethane rigid foam, *J. Cell. Plast.*, 36, 45, 2000.

47. Lamberts, W.M., 1,1,1,4,4,4-hexafluorobutane, a new non-ozone-depleting blowing agent for rigid PUR foams, *J. Cell. Plast.*, 28, 584, 1992.
48. Kenyon, W.G., New, long term alternative fluoro-solvents for electronics cleaning and drying applications, *Trans. IMF*, 74, 142, 1996.
49. Zipfel, L. et al., HFC-365mfc: a versatile blowing agent for rigid polyurethane foams, *J. Cell. Plast.*, 35, 328, 1999.
50. Vachon, C. and Gendron, R., Evaluation of HFC-245fa as an alternative blowing agent for extruded polystyrene, *Cell. Polym.*, 22, 295, 2003.
51. Vachon, C. and Gendron, R., Foaming polystyrene with mixtures of carbon dioxide and HFC-134a, *Cell. Polym.*, 22, 75, 2003.
52. Meyke, J., Berstorff Maschinenbau, H., and Hunziker, P., Extrusion of TPE profiles using water as a physical blowing agent, *Rubber India*, 49, 11, 1997.
53. Sahnoune, A., Foaming of thermoplastic elastomers with water, *J. Cell. Plast.*, 37, 149, 2001.
54. Pallay, J. and Berghmans, H., Water-blown expandable polystyrene: improvement of the compatibility of the water carrier with the polystyrene matrix by in situ grafting, *Cell. Polym.*, 21, 19, 2002.
55. Gendron, R. et al., Foam extrusion of PS blown with a mixture of HFC-134a and isopropanol, *Cell. Polym.*, 23, 1, 2004.
56. Daigneault, L.E. and Gendron, R., Blends of CO_2 and 2-ethyl hexanol as replacement foaming agents for extruded polystyrene, *J. Cell. Plast.*, 37, 262, 2001.
57. Beaujean, M., Methylal: a blowing or co-blowing agent for polyurethane and other foams, in *Proceedings of Blowing Agents and Foaming Processes 2003*, Rapra Technology Limited, Shawbury, U.K., 2003, 25.
58. Modesti, M., Baldoin, N., and Simioni, F., Formic acid as a co-blowing agent in rigid polyurethane foams, *Eur. Polym. J.*, 34, 1233, 1998.
59. Takada, N. et al., Fundamental study of fluorinated ethers as new generation blowing agents, *J. Cell. Plast.*, 35, 389, 1999.
60. Nimitz, J., New foam blowing agents containing fluoroiodocarbons, in *Proceedings of Polyrethanes 1994 Conference*, The Society of the Plastics Industry, Polyurethane Division, New York, 1994, 110.
61. Bogdan, M. and Williams, D., HFC-245fa systems co-blown with CO_2 (water): a quality, cost effective HFC alternative for construction foam applications, in *Proceedings, Polyurethanes 2000 Conference*, American Plastics Council, U.S.A., 2000, 163.
62. Bogdan, M. and Williams, D., A quality, cost effective, flammable HFC blowing agent alternative for construction foam applications: blends of HFC-245fa and hydrocarbons, in *Proceedings Polyurethanes 2000 Conference*, American Plastics Council, U.S.A., 2000, 179.
63. Dournel, P., Zipfel, L., and Berthélemy, P., Optimization and formulations based on HFC-365mfc and blends thereof, *J. Cell. Plast.*, 36, 267, 2000.
64. Zipfel, L. and Boerner, K., Solkane 365/227 blends for polyurethane foaming: skills on commercial use and bulk handling, in *Proceedings of Blowing Agents and Foaming Processes 2003*, Rapra Technology Limited, Shawbury, 2003, 47.
65. Bratt, C. and Albouy, A., Technical and environmental acceptance of HFCs as blowing agents for XPS boards, in *Proceedings of Blowing Agents and Foaming Processes 2001*, Rapra Technology Limited, Shawbury, 2001, paper 15.
66. Albouy, A. et al., Development of HFC blowing agents. Part I: Polyurethane foams, *Cell. Polym.*, 17, 75, 1998.
67. Albouy, A., Guilpain, G., and Crooker, R.M., U.S. Patent 6,624,208, 2003.

68. Miller, L.M. et al., U.S. Patent 6,350,789, 2002.
69. Shutov, F.A., Blowing agents for polymer foams, in *Handbook of Polymeric Foams and Foam Technology*, Klempner, D. and Frisch, K.C., Eds., Hanser Publishers, Munich, 1991, chap. 17.
70. Moulinié, P. et al., Low density foaming of poly(ethylene-co-octene) by injection molding, in *Proceedings of ANTEC 2000*, Society of Plastics Engineers, Orlando, 2000, 1862.
71. Celogen® Physical Properties Guide, Crompton Corporation. www.crompton .com.
72. Gendron, R. and Vachon, C., Effect of viscosity on low density foaming of poly(ethylene-co-octene) resins, *J. Cell. Plast.*, 39, 71, 2003.
73. Luebke, G., Advantages of the combined usage of endothermic and exothermic chemical blowing agents (new developments in the masterbatch technology), in *Proceedings of Blowing Agent Systems: Formulations and Processing*, Rapra Technology Ltd., Shawbury, 1998, paper 11.
74. Champagne, M.F. et al., Foaming polyethylene with CO_2-based mixtures of blowing agents, in *Proceedings of ANTEC 2004*, Society of Plastics Engineers, Chicago, 2004.
75. Moulinié, P., Gendron, R., and Daigneault, L.E., Gas solubility as a guide to physical blowing agent selection, *Cell. Polym.*, 17, 383, 1998.
76. Chaudhary, B.I. and Johns, A.I., Solubilities of nitrogen, isobutane and carbon dioxide in polyethylene, *J. Cell. Plast.*, 34, 312, 1998.
77. Thomas, Y. et al., In-line NIR monitoring of composition and bubble formation in polystyrene/blowing agent mixtures, *J. Cell. Plast.*, 33, 516, 1997.
78. Gendron, R. et al., Foam extrusion of polystyrene blown with HFC-134a, *Cell. Polym.*, 21, 315, 2002.
79. American Society for Testing and Materials, ASTM D1622-03, Standard test method for apparent density of rigid cellular plastics.
80. American Society for Testing and Materials, ASTM D2856-94, Standard test method for open-cell content of rigid cellular plastics by the air pycnometer.
81. American Society for Testing and Materials, ASTM D6226-98e1, Standard test method for open cell content of rigid cellular plastics.
82. American Society for Testing and Materials, ASTM D3576-98, Standard test method for cell size of rigid cellular plastics.
83. American Society for Testing and Materials, ASTM D2842-01, Standard test method for water absorption of rigid cellular plastics.
84. Vachon, C., Evaluation of HFC-245fa as an alternative blowing agent for extruded low-density polyethylene, *Cell. Polym.*, 23, 109, 2004.
85. Gendron, R., Daigneault, L.E., and Caron, L.M., Rheological behaviour of mixtures of polystyrene with HCFC 142b and HFC 134a, *J. Cell. Plast.*, 35, 221, 1999.
86. Chow, T.S., Molecular interpretation of the glass transition temperature of polymer-diluent systems, *Macromolecules*, 13, 362, 1980.
87. Sato, Y. et al., Solubilities and diffusion coefficients of carbon dioxide and nitrogen in polypropylene, high-density polyethylene, and polystyrene under high pressures and temperatures, *Fluid Phase Equil.*, 162, 261, 1999.
88. Gorski, R.A., Ramsey, R.B., and Dishart, K.T., Physical properties of blowing agent polymer systems. 1. Solubility of fluorocarbon blowing agents in thermoplastic resins, *J. Cell. Plast.*, 22, 21, 1986.

89. Park, C.B., Baldwin, D.F., and Suh, N.P., Effect of the pressure drop rate on cell nucleation in continuous processing of microcellular polymers, *Polym. Eng. Sci.*, 35, 432, 1995.
90. Zhang, Z.H. and Handa, P.Y., *In situ* study of plasticization of polymers by high-pressure gases, *J. Polym. Sci. Part B: Polym. Phys.*, 36, 977, 1998.
91. Ettouney, H. and Majeed, U., Permeability functions for pure and mixture gases in silicone rubber and polysulfone membranes: dependence on pressure and composition, *J. Membr. Sci.*, 135, 251, 1997.
92. Arora, K.A., Lesser, A.J., and McCarthy, T.J., Preparation and characterization of microcellular polystyrene foams processed in supercritical carbon dioxide, *Macromolecules*, 31, 4614, 1998.

5

Investigating Foam Processing

Jacques Tatibouët

CONTENTS

0-8493-1701-0/05/$0.00+$1.50
© 2005 by CRC Press

5.1   Introduction: Understanding Extrusion Foam Processsing

The rapidly growing use of polymer foams in various industrial domains results from an increase in the quality of these materials. Their interesting properties, including light weight, high strength to weight ratio, and insulating abilities, have been enhanced through a better understanding of the foaming processes. If a basic understanding of the relationships between the processing conditions and foam morphology (bubble size and foam density) remains important to controlling the cellular structure and producing high quality foams, it is obvious that there is a need to understand the detailed mechanisms at the molecular level. Though predictive models for solubility and plasticization by foaming agent are being developed and improved, a global model able to predict bubble nucleation and growth does not yet exist. This model will be a key step in understanding the fundamentals of complex foam extrusion and in allowing for the development of foamed materials that are well adapted to new applications.

Different characterization methods could be used to investigate the critical steps of the foam extrusion process, where the interaction between the polymer and the blowing agent plays several roles. If specific techniques for separately studying the solubility of blowing agents, the plasticization of polymer melts, and nucleation and bubble growth are numerous, only a few can be used in-line during extrusion.

5.2   Tools for Understanding Foam Extrusion

5.2.1   Introduction

The vast majority of thermoplastic polymer foams are prepared by extrusion. This process involves several key steps in which the nature and the properties of the physical foaming agent (PFA) and its interactions with the polymer play an important role. These steps are, successively, the dissolution of the PFA, the plasticization of the polymer matrix, the nucleation of bubbles upon pressure release, bubble growth, and, finally, cell structure stabilization. The fundamental properties, such as PFA solubility, mixture viscosity, and even interfacial tension need to be investigated, and it is obvious from the above steps that the characterization of all the aspects of PFA properties is not a simple task and involves a great deal of experimentation. Moreover, one would need to make use of several different techniques. These include viscosity measurements and glass transition temperature determinations, as well as direct visual observations or more complex wave scattering experiments. These experimental techniques can be performed off-line, as when a discrete sample is characterized in the laboratory; however, the dynamic aspect of the extrusion process is not taken into account. To assess the influence of the entire set of process parameters, in particular flow or shear conditions, in-line characterization techniques must be operated directly at the process stream and need to be developed.

5.2.2 Solubility

The amount of PFA dissolved in a polymer matrix is not only dependent on the interaction between the polymer and the PFA, but also on the equipment and processing conditions. Solubility remains a function of both pressure and temperature, but during extrusion, sufficient mixing and time is needed to achieve the required concentration of the PFA.

Solubility measurements are usually done in static mode — that is, without any flow or mixing of the melt material. Classical measurements are performed by gravimetric [1] or volumetric methods [2,3]. System pressure and temperature are measured, and the amount of solute gas in the polymer is deduced by a mass balance on the system. As for pressure monitoring [4] or pressure decay techniques [5], these methods are carried out off-line with the system at thermodynamic equilibrium. The time to reach equilibrium may be long especially when the specimen is not limited to powder or thin films. This implies the need for careful calibrations of pressure and temperature sensors. To obtain accurate solubility measurements when a large amount of foaming agent is dissolved, data corrections for swelling, based on the knowledge of equations of states, are needed. The above experimental techniques provide information on the physical chemistry of PFA–polymer systems that is basic to the selection of the appropriate PFA. The data may not be suitable for controlling the extrusion process, due to the fact that equilibrium is not reached in the extruder and that flow and shear induce different solubility limits. A recent work [6] has explored the impact of mixing on gas absorption and the results clearly support the need for solubility data under processing conditions. Near-infrared (NIR) spectroscopy [7,8] can be used in-line to determine the solubility limit with the help of NIR sensors located at the extruder exit. The baseline of the absorption spectra is sensitive to both temperature and to flow rate, and data processing is needed. Other quantitative analysis methods, such as partial least squares (PLS) or principal component regression (PCR), are currently used and give good results on the predictive accuracy of the concentration of PFA in the system. The only limitation may be when fillers are present in the mixture, when the incident light is scattered out and the absorbance may become too weak to be analyzed precisely. But this is not the case when a nucleating agent such as talc is used due to the low concentration generally involved. Optical methods based on direct observation of bubble nucleation in a "transparent" die also demonstrate their ability to obtain dynamic data [9]. On-line rheometry is yet an interesting way to investigate the dynamic solubility properties of blowing agents [10]. This is performed easily by using a "return-to-stream" rheometer and the limit of solubility of PFA is obtained from viscosity measurements performed at a constant stress level.

5.2.3 Plasticization

Dissolution of a low molecular weight PFA in the polymer melt plasticizes the melt matrix and thus leads to viscosity and elasticity reduction. The viscoelastic property spectra are shifted toward low temperatures with increasing PFA concentration enabling foam extrusion to be performed at much lower temperatures than those for the neat polymer. For example, polystyrene/HCFC-142b mixtures are currently processed at temperatures as low as 110–120°C, compared to the classical extrusion temperature of neat polystyrene material around 180–190°C. Low extrusion temperatures are usually one of the prerequisites for preventing bubble coalescence and facilitating processing.

The degree of plasticization can be quantitatively estimated by the lowering of the glass transition temperature of the material. The depression of the glass transition temperature by small molecules or plasticizers can be related to molecular weight, size, and concentration of the diluent through classical and statistical thermodynamics considerations [11]. Under the pressure conditions necessary for maintaining a one-phase mixture, all techniques capable of determining glass transition temperature can be used. Pressurized differential scanning calorimetry (DSC) [12] is one of the preferred techniques, but one difficulty is the possible desorption of gas molecules prior to sealing the sample pans. Deflection temperature measurement [13] has also been used to measure the change in elastic modulus associated with the plasticization.

The viscosity reduction related to the dissolution of PFA is generally explained by an enhancement of chain mobility and is often expressed in terms of change in the free volume. The measurement of viscosity as a function of PFA content is also a way to determine the degree of plasticization or the efficiency of the PFA. This can be performed off-line using a high-pressure capillary rheometer [14,15]. These experiments are not easy to perform and a complex transfer device is needed to introduce the polymer–PFA mixture under pressure in the rheometer. In-line measurements are generally done with the help of a slit rheometer [16,17] or a capillary rheometer [18] attached at the exit of the extruder. Viscosity is derived from classical flow theory — i.e., through the pressure gradient and the volumetric flow rate. The volumetric flow rate is calculated using the measured mass flow rate and the density estimated from an equation of state or from pressure–volume–temperature (PVT) data. Another interesting way to get data is by using an on-line rheometer (a "return-to-stream" type) that permits the collection of viscosity data in a larger shear rate range, directly providing a value for the volumetric flow rate using a gear pump system [10]. In all cases, classical viscoelastic scaling methods, using a composition-dependent shift factor to scale both viscosity and shear rate, lead to master curves, allowing for an estimation of the plasticization efficiency of the PFA. Although this technique does not determine directly the true glass transition temperature (T_g), appropriate models based on the variation of the free volume with respect to the T_g can be used to fit the viscosity data and to provide good estimates indirectly of the shifted glass transition temperatures [10].

5.2.4 Nucleation

The success of the foaming process depends critically on the density of nucleation sites and on the flow behavior inside the extrusion line. If the design and the operating conditions (pressure, temperature, and shear rate) are not adequate for a selected PFA, premature foaming, leading to poor foam quality, is possible. This is dependent on the thermodynamic properties of the foaming agent.

Nucleation can be investigated by two different paths. The first is associated with determination of the processing conditions where phase separation starts and aggregates or clusters of gas molecules collect to form bubble nuclei. This path thus involves techniques that are able to detect phase separation and the early stage of bubble growth, in addition to measurement of pertinent parameters such as temperature, pressure, or shear rate.

Direct visual observation in flow channel or die equipped with a quartz glass window is naturally the most common method. Leading work made by Han and coworkers [19,20] has been carried out and used to determine the pressure and stress conditions where bubbles began to grow. However there has been little study to observe the dynamic behavior of bubble nucleation during foam extrusion. Despite an improvement of this technique by high-speed video cameras [21], the exact position where nucleation occurs is not determined easily and is thus difficult to relate to the shear field.

Light scattering technique can be considered as an indirect method, and it has been successfully used under static or flow conditions [22,23]. With a classical He-Ne laser light source, the light scattering technique allows detection of bubble with a radius as small as 0.5 μm, but this method is restricted to materials without nucleating agents. Phase separation has been detected by more sophisticated techniques, such as time-resolved small-angle X-ray scattering [24] or neutron scattering [25].

The second path for investigating nucleation relies on methods devoted to the determination of the foam morphology due to given processing conditions. These include cell density and size measurements together with foam density. Thus, any imaging technique can be used to characterize samples collected at the die exit. The most common technique involves optical or scanning electron microscopy followed by image processing. More sophisticated techniques such as nuclear magnetic resonance (NMR) imaging [26] or X-ray phase imaging [27] can also be performed for a 3-D imaging purpose.

5.3   The Ultrasonic Technique

Ultrasonic techniques have proven to be useful for investigating elastic solids or viscous fluids, but are not common practice for the study of polymers. However, in the last decade, these nondestructive methods were adapted to polymer characterization and can be now easily used to monitor polymer processing.

Ultrasonic techniques measure the propagation characteristics of small amplitude mechanical waves through the polymer. The wave propagation velocity and the attenuation are related to the physical properties of the material such as viscosity, composition, occurrence of a second phase, and the presence of inclusions or bubbles in the medium. The technique can be operated either off-line (no flow conditions) or in-line during extrusion. Hence, in addition to being non-invasive, this technique does not change the material's properties.

5.3.1   Basics: The Propagation of Sound Waves in Polymers

Ultrasonic techniques for measuring elastic or viscous polymer properties require the use of high frequencies in the range of 0.5 to 20 MHz. Low amplitude mechanical waves operates at stress levels of $\sigma = 10^2$ Pa, strain levels of $\varepsilon = 10^{-7}$, and amplitude of 0.01 nm. At a level of ~1 μW/cm^2, the energy involved is so low that the system is not significantly perturbed.

The propagation of a longitudinal wave into a polymer is governed by the density and by the complex longitudinal modulus of the material $L = L' + iL''$, resulting from a linear combination of the bulk (K) and the shear (G) modulus, $L = K + 4G/3$. The ultrasonic velocity V_{us} for the longitudinal wave is given by the following:

$$V_{us} = (L'/\rho)^{1/2} = [(K' + 4G'/3)/\rho]^{1/2} \tag{5.1}$$

where ρ is the density of the material. For polymers, the contribution of the shear modulus to the storage modulus (L') can be neglected in comparison with that of the bulk modulus ($G << K$), so Equation 5.1 can be rewritten using the hydrostatic compressibility, $\kappa \equiv 1/K'$:

$$V_{us} = (K'/\rho)^{1/2} = (\rho\kappa)^{1/2} \tag{5.2}$$

Attenuation for a homogeneous fluid describes the effect of viscous losses and is given by the following:

$$\alpha_\eta = (\eta_{us}\omega^2)/(2\rho V_{us}^3) \tag{5.3}$$

where ω is the angular frequency ($\omega = 2\pi f$). It should be noted that η_{US} in Equation 5.3 is defined as $\eta_{US} = L''/\omega$ and differs from the viscosity, η, measured through classical rheological experiments.

In these experiments, macroscopic displacements in the order of μm are involved, whereby chains reptate as a whole through the entanglement network. By comparison, the ultrasonic wave is a very small disturbance with typical magnitude below the nanometric range, which, in any case, cannot dismantle entanglements. Therefore, ultrasonic viscosity η_{US} relates to the small-scale mobility of short segments between the entanglements. While η is mainly associated with molecular weight, polydispersity, and long chain branching, η_{US} is governed by the chemical nature and morphology of the repeating units and by the distance between entanglements.

For heterogeneous materials, such as filled polymers, polymer blends, composites, and polymeric foams (in the bubble nucleation phase), acoustic waves are subjected to other sources of attenuation, which will be denoted as α_S. This additional attenuation is generated by two primary mechanisms: scattering and viscous drag.

Scattering is observed when the wavelength is of the same order of magnitude as the diameter of the foreign particles. This generally can be the case when high frequencies are used or when particles have a large diameter and the mechanism is subjected to contrast in density and bulk modulus between the two media (fluid and particles) [28–31].

Other factors include the displacement of the particles in the fluid when exposed to the acoustic wave. This mode of dissipation prevails when low frequencies are used or small particles are present [32,33]. The viscous drag is a function of the projected area of the particle.

5.3.2 Off-Line Technique

An original ultrasonic characterization device, which mimics the temperature and pressure conditions that prevail in most of the polymer processes, has been used for materials research purposes in amorphous polymers [35,36] subjected to pressure and temperature sweeps in addition to studies of semicrystalline polymers [37,38].

As illustrated in Figure 5.1(a), the polymer is held in confinement between two axially aligned steel buffer rods, at the end of which the ultrasonic transducers are mounted. The longitudinal waves are produced with piezoelectric transducers having a center frequency of 2.25 MHz. The emitting transducer is repetitiously energized approximately every 0.01 s to produce a short burst of ultrasounds, lasting typically 2 to 5 periods (Figure 5.1[b]). The acoustic pulses with amplitude A_o travel down to the steel/polymer interface where part of the energy is transmitted to the polymer sample of thickness e. The acoustic stress is then reflected back and forth between the two interfaces, its energy being decreased at each reflection. As a function of time this produces a series of echo signals, A_1, A_2..., exiting from the

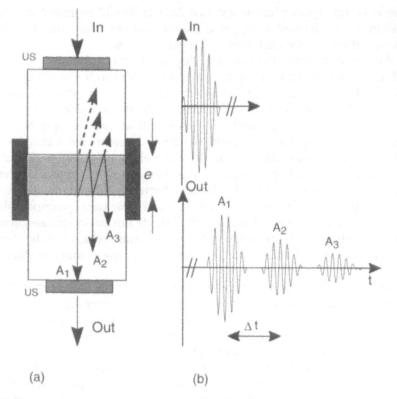

FIGURE 5.1
Basic principle of ultrasonic measurement: (a) Setup; (b) echo signals from emitting transducer (IN), detected by receiving transducer (OUT). (From Sahnoune, A. et al., *J. Cell. Plast.*, 37, 429, 2001 [52]. With permission.)

second interface and directed toward the receiving transducer (Figure 5.1[b]). Practically, only the first two echoes are sent to a broadband amplifier connected to a fast digitizer. The digital signals are then processed using a computer in addition to fast Fourier transforms (FFT) and cross-correlation techniques. The sound velocity V_{US} can be found by measuring the time delay Δt between successive echoes, as represented by Equation 5.4:

$$V_{US} = 2e/\Delta t \tag{5.4}$$

The attenuation α is determined from the ratio of the amplitudes of successive echoes and is defined on a logarithmic scale in dB/cm:

$$\alpha = -10 \log\left[\left(1/\Gamma^2\right)\left(A_2/A_1\right)\right]/e \tag{5.5}$$

with Γ being the reflection coefficient at the polymer–buffer rod interface.

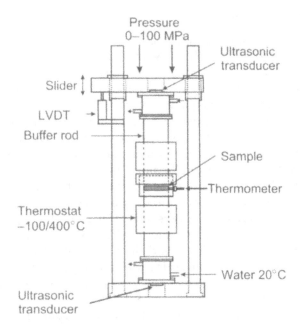

FIGURE 5.2
Schematic of the off-line (no flow) ultrasonic measurement device. (From Sahnoune, A. et al., *J. Cell. Plast.*, 37, 429, 2001 [52]. With permission.)

Figure 5.2 is a schematic drawing of the off-line ultrasonic apparatus [34]. The sample material is confined between two steel rods, by means of a sleeve. The upper rod moves to accommodate changes in sample thickness e, which is continuously monitored to ± 1 µm for the determination of specific volume (V) and density $(\rho = 1/V)$. The system controls a clamping pressure up to $P = 100$ MPa, temperature T between -100 and $400°C$, and heating and cooling rates from $50°C/min$ to $1°C/h$. Moreover, in isothermal measurements, the applied pressure can be easily varied at different rates from 0.01 up to 10 MPa/min. Measurements can be performed during temperature or pressure sweeps, or as a function of time in the case of kinetic studies. Typically, these experiments use 3 to 4 g of polymer.

In the case of foam processing studies, two methods can be used to prepare the polymer–PFA mixture samples. The first one is an *in situ* approach where the PFA is injected at high pressure into the confined molten polymer. This requires supplying the PFA directly into the sample holder of the ultrasonic device. The second method uses a separate pressure vessel where PFA is dissolved in the polymer sample. After quenching and pressure removal, the sample is quickly transferred to the ultrasonic apparatus. Both preparation methods give similar results. The precise concentration of PFA is later determined by weighting the sample before and after degassing under vacuum inside a high temperature oven.

5.3.3   In-Line Technique

The static setup (off-line) was adapted for in-line monitoring during extrusion, using the instrumented slit die shown in Figure 5.3 [39]. Two identical ultrasonic probes were installed at midstream in the die and were set perpendicular to the slit, so that the ultrasonic beam was normal to the flow. Longitudinal waves at a frequency of 5 MHz were used. The slit die, mounted at the end of the extruder, had a channel thickness and width of 4.5 and 40 mm, respectively. The die was also equipped with a flush-mounted thermocouple to measure the melt temperature and three pressure transducers to measure the pressure profile along the die. Pressure measurements allow calculation of the stress level in the die and the pressure at the ultrasound probes location. The electronics for generation and detection of the ultrasound were contained in a personal computer. The signals were digitized, and the ultrasonic parameters (attenuation and velocity) were extracted using specialized high-speed signal analysis algorithms. The computer also acquired the data concerning the process parameters, temperature, and pressure. The information was analyzed and the results were provided to the user on a real-time basis.

Numerous works specifically aimed at industrial applications [40–46] demonstrate that this technique is a powerful tool for process monitoring.

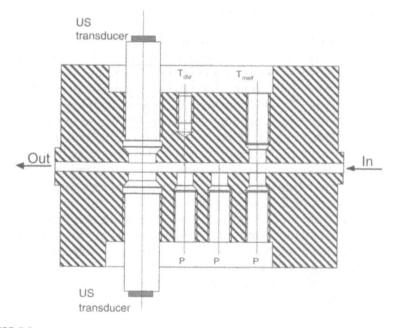

FIGURE 5.3
Instrumented slit die used for in-line ultrasonic characterization; P indicates the location of pressure sensors and T_{melt} the location of the melt thermocouple. (From Sahnoune, A. et al., *J. Cell. Plast.*, 37, 429, 2001 [52]. With permission.)

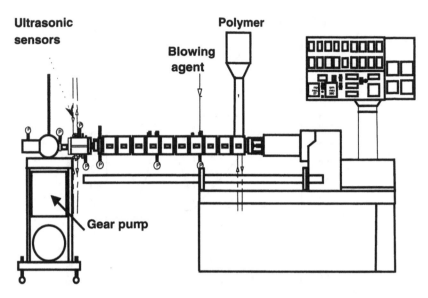

FIGURE 5.4
Foam extrusion setup using twin-screw extruder, instrumented die, and gear pump. (From Tatibouët, J. et al., *J. Cell. Plast.*, 38, 203, 2002 [50]. With permission.)

In experiments involving foam extrusion, the instrumented channel was attached to the exit of a twin-screw extruder. A gear pump, located after the slit die, allowed for the control of the pressure inside the flow channel (Figure 5.4).

5.4 Applications of Ultrasounds Related to Foam Processing

The recently developed ultrasonic technique, which can be used to investigate all stages of polymer foaming, is illustrated through several examples from work carried out over the past few years at the Industrial Materials Institute of the National Research Council of Canada, mainly on foam extrusion of polystyrene with different physical blowing agents such as CO_2, HCFC-142b, and HFC-134a. The last example is related to chemical foaming of crosslinked polyolefins. Application of the ultrasonic method to investigate nucleation is also covered.

5.4.1 Solubility

Solubility and diffusivity of the gas in the polymer matrix are key parameters in foam processing because they are the limiting factors in determining the foam density and also in regulating the phase separation dynamics during

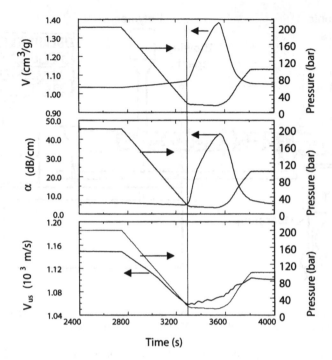

FIGURE 5.5
Off-line experiment with controlled pressure sweep for a mixture of polystyrene with 1.7 wt% of CO_2 at 180°C. Top: specific volume; middle: attenuation; bottom: ultrasonic velocity. Vertical line at $t \approx 3300$ s indicates onset of phase separation. (From Sahnoune, A. et al., *J. Cell. Plast.*, 37, 429, 2001 [52]. With permission.)

bubble growth. As mentioned before, people agree generally on the two terms in reference to a static mode in thermodynamic equilibrium. The ultrasonic technique provides an original way to study phase separation dynamics of the mixture polymer–PFA and cell nucleation kinetics. The ultrasonic off-line apparatus is gas tight and by reducing the applied pressure one can induce foaming at a given temperature without a loss of foaming agent. Figure 5.5 illustrates the results of a typical experiment. Specific volume, attenuation, and velocity show drastic changes when the critical foaming pressure (phase separation) is reached. The procedure can be repeated at different temperatures and pressure drop rates. It is therefore possible to determine the critical foaming pressure and the onset of phase separation under various conditions. The reverse process is also possible, and dissolution of the foaming agent when the pressure is increased again can be studied.

However, it has been frequently reported that shear has an influence on bubble nucleation [47] and in-line measurements are certainly the best way to determine the degassing conditions in terms of temperature and pressure prevailing at the location of the ultrasonic probes, thus establishing a steady flow of homogeneous mixture. The pressure in the instrumented channel is

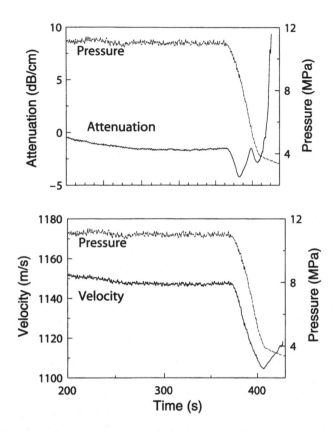

FIGURE 5.6
In-line measurement of ultrasonic properties. Top: attenuation; bottom: ultrasonic velocity
during pressure reduction for a mixture of polystyrene with 11 wt% HCFC-142b at 125°C. (From
Sahnoune, A. et al., *J. Cell. Plast.*, 37, 429, 2001 [52]. With permission.)

maintained at a high enough level to prevent phase separation. In a second
step, the pressure is lowered rapidly at a rate of approximately 200 kPa/s
by steadily increasing the speed of the gear pump. During these operations,
the pressures in the slit channel and the ultrasonic parameters (attenuation
and velocity) are continuously monitored. The degassing pressure is generally defined as either the point where the attenuation rises (Figure 5.6), which
is related to the scattering of the ultrasonic wave as the size and the number
of gas bubbles increase, or the point where velocity reaches the minimum,
which is associated with the onset of phase separation [47].

5.4.1.1   Apparent Effect of Flow on Degassing

A comparison between quasi-static conditions (off-line) and dynamic degassing (in-line) is illustrated in Figure 5.7. Bubble nucleation takes place more
readily when the mixture is exposed to a shear field, the stresses induced

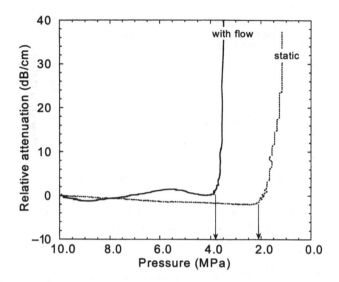

FIGURE 5.7
Comparison between static (no flow) and dynamic degassing for mixture of polystyrene with 12 wt% of HCFC-142b at 135°C. Arrows indicate the degassing pressure. Note that x-axis is reversed. (From Sahnoune, A. et al., *J. Cell. Plast.*, 37, 429, 2001 [52]. With permission.)

by the flow conditions lowering the energy barrier between the stable solution state and the unstable bubble phase.

For given content of PFA and shear rate, it is possible to establish characteristic curves where degassing pressure is plotted as a function of temperature. An interesting feature of these curves is certainly their parabolic shape, as shown in Figure 5.8. The minimum delimits two domains. In the high temperature part, the melt viscosity is low and plays a minor role in the foaming process. In this region, the blowing agent solubility alone, as determined under static conditions, controls the phase separation process. Under these conditions, there is a fast growth of the cells, which in turn results in cells collapsing and/or coalescence with a large part of the gas escaping through the cell walls. On the other hand, in the low temperature region, the melt has a higher viscosity (and non-negligible elasticity), a necessary condition for good foaming. The increase in the measured degassing pressure in this region is a reflection of the viscoelastic response of the polymer, the pressure drop rate, and the influence of flow. The elastic response of the polymer plays an important role as the characteristic relaxation time of the materials becomes comparable to that of the extrusion experiment [48]. This aspect of viscosity-dependent nucleation will be detailed in a further paragraph.

5.4.1.2 Apparent Effect of Nucleating Agent on Degassing

The use of nucleating agents is in most cases necessary to increase the cell density and to modify the dynamics of bubble formation and growth [49].

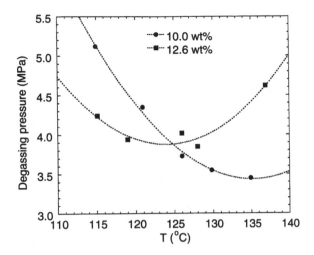

FIGURE 5.8
Degassing pressure as a function of temperature for two concentrations of HCFC-142b in polystyrene. (From Sahnoune, A. et al., *J. Cell. Plast.*, 37, 429, 2001 [52]. With permission.)

In the case of semicrystalline polymer matrices, nucleating agents may also impact the crystallization process.

Figure 5.9 shows that the degassing pressures for the mixtures of PS/HCFC-142b with nucleating agents are higher than those with neat PS over the whole range of temperature. As one can observe from the same figure, TiO_2 and talc are clearly discriminated, and the arising question addresses the nucleating conditions.

The volume concentrations of the nucleating agent are 0.48 and 0.7 vol% for TiO_2 and talc, respectively, but due to the mean particle size, the number of particles per unit volume is considerably higher in the case of TiO_2. In principle, a higher number of potential nucleation sites is expected to lead to a higher cell density, as described by classical heterogeneous nucleation theory. This is clearly not the case, as was observed from the foam morphology. On the contrary, talc leads to a much finer structure than that using TiO_2.

The higher degassing pressures found in mixtures with a nucleating agent are to be compared with the result for 12 wt% of HCFC-142b at 180°C, as indicated in Figure 5.9. A 2% increase in concentration of blowing agent leads to a 1.2 MPa increase of the degassing pressure for neat PS mixture, similar to the increases measured for talc and PS with 10 wt% HCFC-142b. This suggests that the local gas concentration in the mixture would be higher than the average nominal value of 10 wt%. This higher concentration can be located close to the interfaces between polymer and nucleating agent particles, and ultimately these locations become nucleating sites as soon as the pressure reaches the value corresponding to higher concentrations [50]. Depending on the nature of the nucleating agent, interaction with gas molecules will be different and the local concentration of HCFC-142b at the interfaces is expected to vary. The thermodynamic properties, especially

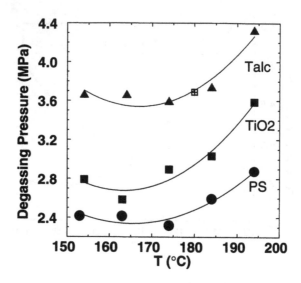

FIGURE 5.9

Degassing pressure as a function of temperature for three mixtures based on polystyrene with 10 wt% HCFC-142b, with and without nucleating agent. TiO_2 and talc concentration are set at 2 wt%. The open-square symbol represents the degassing pressure for a neat polystyrene mixture with 12 wt% of HCFC-142b at 180°C. (From Tatibouët, J. et al., *J. Cell. Plast.*, 38, 203, 2002 [50]. With permission.)

the surface energy of the nucleating agents, would in this case play a significant role.

5.4.2 Plasticization

When low molecular weight PFA molecules are dissolved under pressure in the polymer melt, plasticization occurs and hence leads to viscosity and elasticity reduction. This plasticization enables foam extrusion to be performed at low temperatures close to the glass transition temperature of the neat polymer. As mentioned earlier, this aids in extruded foam stabilization (low extrusion temperatures prevent bubble coalescence) and facilitates processing [51]. The ultrasonic technique allows either off-line or in-line quantification of the amount of plasticization induced by the foaming agent.

Off-line characterization gives a direct access to glass transition temperature. Figure 5.10 shows the plasticization effect of 1.7% CO_2 dissolved in PS. Attenuation and velocity curves are both shifted toward low temperatures. The glass transition temperature can be determined from the break in specific volume variation, according to the usual definition. This point coincides also with a break in the slope of velocity curves and with the onset of the attenuation peak as shown in the Figure 5.10. T_g, measured either from the ultrasonic properties or from the specific volume, decreases from 102 to 87°C for the tested case. It is important to note that this measurement is done at 20.0 MPa, which is not accessible to classical methods such as DSC or rheology.

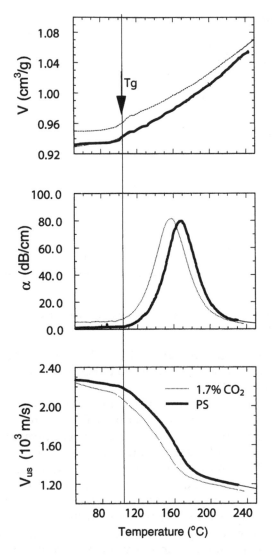

FIGURE 5.10

Off-line experiment showing the shift of properties toward lower temperature when 1.7 wt% CO_2 is dissolved in polystyrene, with pressure set at 20 MPa. The glass transition temperature is defined as (1) changes in the slopes of either specific volume (top) or ultrasonic velocity (bottom) or (2) onset of the relaxation peak (attenuation, middle plot). (From Sahnoune, A. et al., *J. Cell. Plast.*, 37, 429, 2001 [52]. With permission.)

Figure 5.11 presents the glass transition temperature determined by off-line measurements of the velocity of sound and the attenuation as a function of foaming agent concentration for three different foaming agents in polystyrene. These curves are very useful in the choice of a foaming agent–polymer system by demonstrating the efficiency of a blowing agent on a weight basis in lowering the glass transition temperature. As an example, CO_2

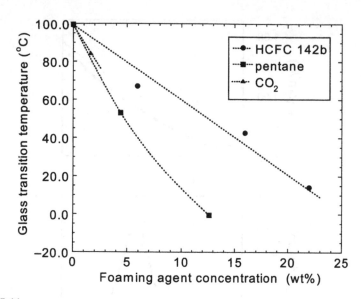

FIGURE 5.11

Glass transition temperature as a function of concentration for different blowing agents in polystyrene. (From Sahnoune, A. et al., *J. Cell. Plast.*, 37, 429, 2001 [52]. With permission.)

induces better plasticization than HCFC-142b. It follows that a lower concentration of carbon dioxide would be required to match the same degree of plasticization induced by HCFC-142b.

Figure 5.12 shows the velocity of sound in the polystyrene/HCFC-142b mixture during extrusion (in-line experiments), as a function of temperature at different concentrations. The data were taken at pressures ranging from 10.3 to 11.0 MPa where phase separation and foaming cannot take place. The velocity of sound is almost half that of neat polystyrene at the same temperature and at the same pressure ($V_{US} \approx 2000$ m/s at 120°C and 10 MPa). This indicates a decrease in intermolecular forces caused by the presence of small gas molecules acting as a plasticizer between the large polymer chain segments. Note that the velocity decreases linearly as the mixture becomes less viscous at high temperatures. Sound velocity is very sensitive to blowing agent concentration: over the concentration range considered, it changes by about 15 m/s for an increment of 1% in HCFC-142b, well above the precision of the measurement technique (0.5 m/s). It is also much larger than the fluctuations that can be caused by variations in pressure or temperature (about 4 m/s/MPa and 2 m/s/°C, respectively [36]). The velocity of sound is therefore an important feature, which provides a simple and unique way to monitor in-line composition of polymer–PFA mixtures.

In addition to the above, data in Figure 5.12 can be used to model glass transition variation with foaming agent concentration in much the same way as that used in rheometry [10]. Typically, corresponding state principles are assumed, and curve shifting is performed at constant ultrasonic velocities.

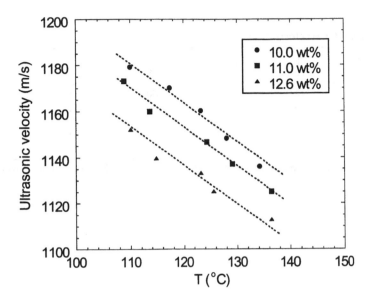

FIGURE 5.12
Ultrasonic velocity as a function of temperature for different HCFC-142b concentrations measured during extrusion. Pressure is maintained constant at 10.3 MPa. (From Sahnoune, A. et al., *J. Cell. Plast.*, 37, 429, 2001 [52]. With permission.)

Shift in temperature due to the added amount of foaming agent corresponds thus to an equivalent shift in the glass transition temperature.

Knowledge of the glass transition temperature for the mixture of polymer and blowing agent is the key element for generating a comprehensive description of the rheological behavior. Thus, through the use of an adequate model [10], reduced viscosity data of the polymer–PFA mixture can be generated from glass transition temperatures provided by the ultrasonic characterization.

5.4.3 Nucleation

5.4.3.1 *Effect of the Rheological Properties of the Polymer*

As mentioned in a previous paragraph, for a given blowing agent content, the curves where degassing pressures are plotted as a function of temperature exhibit a parabolic shape [52,53]. This behavior is linked to the rheological properties of the polymer, as demonstrated in a study based on five linear polystyrene resins with melt flow indices (MFI) varying from 20.3 to 1.6 dg/min (Table 5.1), in which 6 wt% of HFC-134a was dissolved, with no addition of any nucleating agent [48].

As shown in Figure 5.13, the in-line degassing pressures for the low viscosity resins (11.5 and 20.3 MFI) are in good agreement with solubility data

TABLE 5.1

Rheological Characterization of the PS Resins

Resin MFI (dg/min)	η_0 (kPa·s)	$J_e^0 \times 10^5$ (kPa^{-1})	τ_m (s) 155°C	τ_m (s) 165°C	τ_m (s) 175°C
1.6	172.0	13.6	500	200	100
2.3	128.9	13.4	325	135	65
5.3	41.7	15.6	100	40	20
11.5	23.5	9.6	35	15	7
20.3	9.8	8.6	6	3	1

Source: Tatibouët, J. and Gendron, R., *J. Cell. Plast.*, 40, 27, 2004 [48]. With permission.

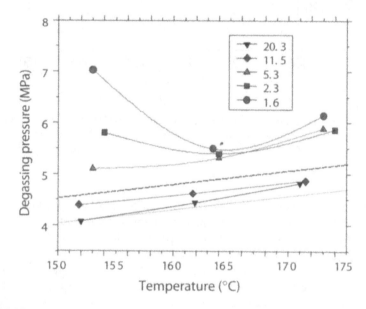

FIGURE 5.13

Degassing pressures as a function of temperature for polystyrene resins of various melt flow indices (MFIs). Resins are identified by their different MFI values. This nomenclature will be kept for the other related figures. Dotted lines correspond to equilibrium solubility data from Sato, Y. et al., *Polym. Eng. Sci.*, 40, 1369, 2000 [3] and Daigneault, L.E. et al., *J. Cell. Plast.*, 34, 219, 1998 [4]. The HFC-134a concentration is set to 6 wt%. (From Tatibouët, J. and Gendron, R., *J. Cell. Plast.*, 40, 27, 2004 [48]. With permission.)

excerpted from Sato et al. [3] and Daigneault et al. [4], over the entire temperature range investigated. A different behavior has been reported, however, for the more viscous resins. At high temperature, their degassing pressures are at least 0.7 MPa above those of the low-viscosity group, and an upward tendency is observed at temperatures lower than 160°C, with the magnitude being proportional to the viscosity of the resin. The most viscous PS (1.6 MFI) even exhibits a degassing pressure of 7.0 MPa at 153°C, which is approximately 2.7 MPa higher than the solubility curve.

The splitting of the resins into two groups at an elevated temperature was associated with their respective elasticity, as reported in Table 5.1 through their value of steady-state compliance J_e^0. This rheological parameter, which is related to the amount of strain recovery or recoil that occurs after the stress is removed, can be estimated from the following [54]:

$$J_e^0 = \frac{G'}{\eta_0^2 \omega^2} \tag{5.6}$$

with G' being the elastic modulus (extrapolated at the lower limit of measuring frequency), η_0, the zero-shear viscosity and ω, the frequency. Low elastic recovery is then associated with the low-viscosity resins that fit the static solubility data, while higher degassing pressures are observed for resins with corresponding larger J_e^0 values.

One manifestation of the elastic recovery of most extruded viscoelastic materials is the extrudate swell (ratio of the extrudate diameter to that of the capillary orifice), which should not be confounded with cellular expansion. Large swell is usually observed with short dies, and increasing the length of the die yields an asymptotic value of the swell. A tensile stress, originating from the extensional deformation that occurs at the inlet of the die, relaxes gradually in the die [56] according to the Weissenberg number W_e, which can be defined as the ratio of the longest relaxation time to that of the residence time within the die (reciprocal of the deformation rate). The remaining tensile stress reaches an equilibrium value set by the first normal stress difference N_1 induced by the shear [57]:

$$N_1 = \sigma_{11} - \sigma_{22} = 2J_e^0 \sigma_{12}^2 \tag{5.7}$$

This tensile stress is maintained in equilibrium until the polymer melt finally relaxes after it exits the die, and corresponds to what is usually referred as the exit pressure, as illustrated in Figure 5.14. Thus, for short dies, or at a short distance with respect to the entrance, the remaining tensile stress may be pretty large.

Given that high viscosity resins will experience higher shear stresses as illustrated in Figure 5.15, and assuming that at high temperatures the tensile stress has relaxed to the equilibrium condition, normal stresses of a few negligible kPa for the 20.3 MFI resin, but a significant 800 kPa for the high viscosity 1.6 MFI resin, are obtained using Equation 5.7. This 800 kPa value for the tensile stress prevailing in the flow axis corresponds to the difference observed in the degassing pressures between the two resins groups at temperatures above 160°C. At lower temperatures, it was assumed that the equilibrium for the relaxation of the tensile stress was not reached yet, especially for the more viscous resins, according to the estimated increasing maximum relaxation time τ_m (Table 5.1). These larger tensile stresses would

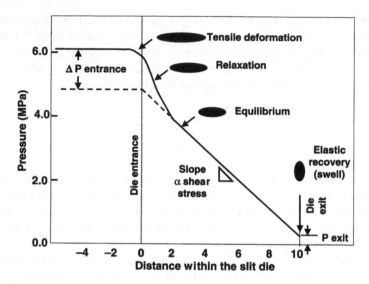

FIGURE 5.14
Evolution of the pressure in a slit die, with corresponding elastic recovery behavior of the melt. (From Tatibouët, J. and Gendron, R., *J. Cell. Plast.*, 40, 27, 2004 [48]. With permission.)

contribute to the reported higher degassing pressures. These findings could also explain the parabolic degassing profile visually observed by Han and Han [22], with the first bubbles to appear being located in the center of the flow, because of the "strong stretching along the centerline" [58]. The elliptical shape of the bubbles, frequently observed during their flow in a slit die, could also be explained by the smaller resistance in the machine direction exerted on the growing cells.

In addition to being sensitive to the phase-separation conditions, the ultrasonic properties as measured in-line can also be linked to the rheological characteristics of the melt, under shear and/or extensional field. Static (no flow) ultrasonic investigation cannot discriminate between different molecular weights or molecular weight distributions, as illustrated in Figure 5.16. However, the local arrangements of the chains (molecular alignment and disentanglement) induced by shear and extensional deformations under flowing conditions, which vary according to the rheological behavior of each resin, have a significant impact on the ultrasonic velocity, as shown in Figure 5.17 for the different PS resins (with the ultrasonic velocity measured prior the degassing ramp, and expressed at constant pressure), or more specifically in terms of the relevant shear stress experienced at low temperature (152°C), as illustrated through a linear dependency in the insert in Figure 5.17. A correlation that would take into account the induced tensile stress can also be attempted, through an estimated stress (efficient pressure) in the machine direction (also schematically represented at the bottom of Figure 5.18):

$$P_{efficient} = P_{hydrostatic} - Tensile\ stress = P_{measured} - (P_{degassing} - P_{solubility}) \quad (5.8)$$

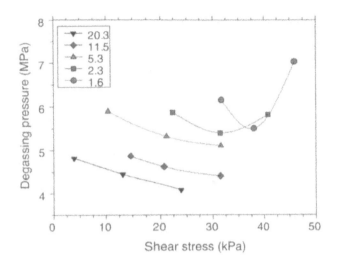

FIGURE 5.15
Degassing pressure as a function of the apparent shear stress. (From Tatibouët, J. and Gendron, R., *J. Cell. Plast.*, 40, 27, 2004 [48]. With permission.)

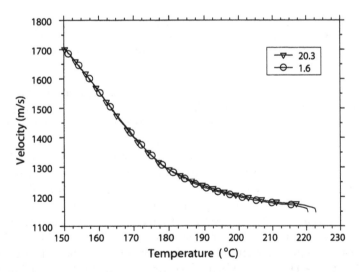

FIGURE 5.16
Sound velocity as a function of temperature, measured under static conditions at 10 MPa for resins with melt flow indices of 20.3 and 1.6. (From Tatibouët, J. and Gendron, R., *J. Cell. Plast.*, 40, 27, 2004 [48]. With permission.)

with the measured pressure $P_{measured}$ being the hydrostatic pressure before the pressure decrease (7.6 MPa). The tensile stress, assumed to be responsible for the reported deviations from the equilibrium solubility data, was thus estimated from the difference between the measured degassing pressure and the pressure associated with static solubility measurements. As plotted in

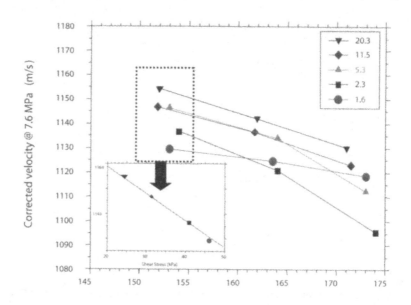

FIGURE 5.17
Sound velocity measured in the homogeneous single phase (and corrected for $P = 7.6$ MPa) as a function of temperature. Insert shows the sound velocity measured at 152°C as a function of the apparent shear stress. (From Tatibouët, J. and Gendron, R., *J. Cell. Plast.*, 40, 27, 2004 [48]. With permission.)

Figure 5.18, the ultrasonic velocity exhibits a linear dependency with respect to this efficient pressure, with a slope of 7.8 m/s/Mpa, which is in close agreement with the pressure dependency of the ultrasonic velocity for PS in the molten state (roughly 6.3–9.3 m/s/MPa).

The fact that the ultrasonic velocity could be an unbiased estimator of the efficient pressure that controls the solubility, could be further validated by plotting the velocity as measured at the degassing point, for all the PS resins, as depicted in Figure 5.19. A single dependency is almost observed, disregarding the sensitivity of ultrasonic velocity to pressure: for the degassing points recorded near 155°C, the pressure spanned between 4.0 and 7.2 MPa for the various resins. In addition, for this temperature the data are grouped close to the value 1130 m/s, which also corresponds to the "true" ultrasonic velocity at 4.0 MPa (equilibrium pressure for solubility of 6 wt% of HFC-134a at 155°C). Thus the ultrasonic velocity, being a function of both elasticity and free volume (see Equation 5.2), would represent adequately the equilibrium conditions satisfying both solubility and rheology.

5.4.3.2 Heterogeneous Nucleation

The addition of nucleating agents to the polymer–blowing agent mixture, by promoting heterogeneous nucleation, increases the nucleation rate, which

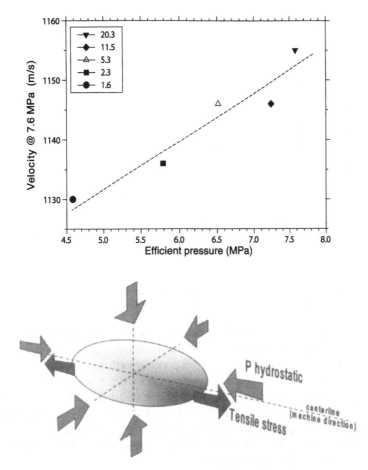

FIGURE 5.18
Top: Sound velocity measured at 7.6 MPa and 155°C as a function of the efficient pressure P_{eff} (Equation 5.8). Bottom: Schematic of the stress components exerted on the flowing melt inside the melt. (From Tatibouët, J. and Gendron, R., *J. Cell. Plast.*, 40, 27, 2004 [48]. With permission.)

enables a better control of the cell morphology (cell density, cell size, and cell size distribution) [59]. The influence of the nature and of the concentration of nucleating agents on cell size and cell size distribution have been investigated in various works [49,60], and it was found that some nucleating agents are more effective than others. In addition, the effect of filler size on cell nucleation interacts with the PFA concentration: while fine particles yields high cell density at high saturation pressures, efficiency is lowered as the saturation pressure is decreased [61].

In complement (or contradiction) to the classical heterogeneous nucleation theory based on the reduction of the interfacial energy with the bubble nuclei being formed at the interface of the polymer melt and that of the particle surface, several other alternative mechanisms have been proposed. For example, undissolved gas can remain in micropores present on the surface

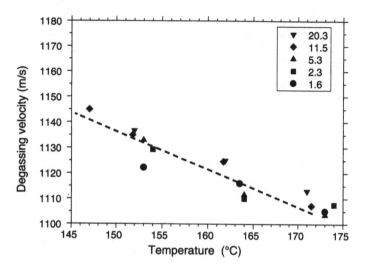

FIGURE 5.19

Sound velocity measured at the onset of degassing as a function of temperature. (From Tatibouët, J. and Gendron, R., *J. Cell. Plast.*, 40, 27, 2004 [48]. With permission.)

of the particles, or can accumulate at the polymer–filler interface. Attempts to validate this last assumption consisted in observing an increase of the gas absorption in filled polymer [6]. These observations are unfortunately in contradiction with another study performed under well-controlled conditions [62], where no modification of the gas solubility in presence of filler was reported. Other models are based on pre-existing microvoids [63] or microcrevices at the nucleating agent surface with the nucleation effect being amplified by shear [64]. However, most of these models ignore the thermodynamics of the multiphase systems.

As reported previously, the ultrasonic method used in-line, in the presence of nucleating particles, has indicated the increase of the degassing pressures with respect to the solubility controlled mainly by temperature and pressure, for a given PFA content. This effect was further explored for PS foaming, by varying the concentrations of talc and HFC-134a and reporting nucleation cell density and degassing conditions (Figures 5.20 and 5.21).

While the degassing pressures are in fair accordance with static solubility results [3], a significant increase in the degassing pressures is observed with the addition of only 1 wt% of talc, with the biggest impact observed at a lower PFA content. Additional step increases of talc had little effect at 6 wt% of HFC-134a. However, for 3 wt% of HFC-134a, increasing the talc content to 3 wt% induced a large jump in the degassing pressure, to a level close to those observed with 6 wt% of HFC-134a. The cell densities, obtained from quantitative analysis of micrographs of the foams, are displayed in Figure 5.21, and exhibit similar trends as those observed for the degassing pressures, with an exception for the abnormally high degassing pressures obtained at 3 wt% of HFC-134a and 3 wt% of talc.

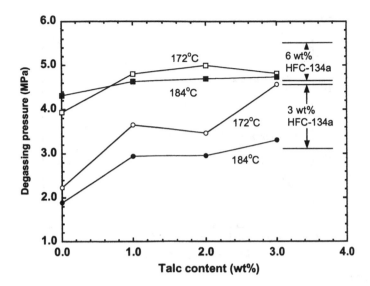

FIGURE 5.20
Degassing pressures for different talc contents, at two HFC-134a concentrations and two temperatures. (From Tatibouët, J. and Gendron, R., *Proc. SPE Annu. Mtg. ANTEC*, 2552, 2004 [77]. With permission.)

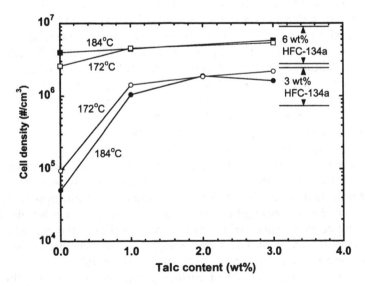

FIGURE 5.21
Cell density obtained under various conditions (talc concentration, HFC-134a concentration, and temperature). (From Tatibouët, J. and Gendron, R., *Proc. SPE Annu. Mtg. ANTEC*, 2552, 2004 [77]. With permission.)

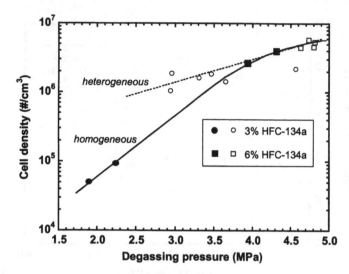

FIGURE 5.22

Cell density (for homogeneous and heterogeneous nucleation) vs. degassing pressure for two HFC-134a concentrations. (From Tatibouët, J. and Gendron, R., *Proc. SPE Annu. Mtg. ANTEC*, 2552, 2004 [77]. With permission.)

Figure 5.22 displays the cell density as a function of the degassing pressure, measured with the ultrasonic probes. The homogeneous nucleation (no talc), indicated by a solid curve, reflects a simple solubility/concentration dependence [55]. The heterogeneous nucleation is also displayed through a master curve (dotted line), which intersects its homogeneous counterpart at a degassing pressure close to the critical pressure of HFC-134a (4.06 MPa). Although classical nucleation theory could explain the high nucleation density values obtained with 6 wt% of HFC-134a, it is surprising to find that the results obtained with talc lie in the same cell density range, while heterogeneous nucleation is expected to be controlled by the number of particles, with maximum efficiency being one cell per particle.

The former hypothesis that the increase of the degassing pressures comes from an increase of the foaming agent concentration localized in the neighborhood of the solid particles [50] necessitates thermodynamical considerations. Minimization of the system's free energy for multiple-component systems may imply segregation of the species even in a miscible system. Maintaining certain regions of the system with a different composition than that of the bulk obviously has a free energy cost, which can be minimized, however, by avoiding steep concentration gradients [65].

Because of the large amount of surface generated by the talc, the plasticizing PFA molecules need to be dragged away from the neighborhood of the solid particle to minimize the difference in the surface energy between the nucleating agent surface (41.6 mN/m for talc [66]) and the PS matrix. In presence of a dissolved gas, the value for neat PS (42 mN/m) can be decreased to as low as 5–10 mN/m, as reported in Jaeger et al. [67] for the

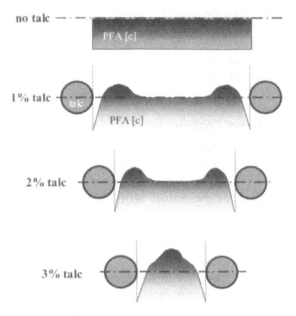

FIGURE 5.23
Schematic model for the PFA gradient in the polymer matrix near the talc particles, as a function of content of talc. (From Tatibouët, J. and Gendron, R., *Proc. SPE Annu. Mtg. ANTEC*, 2552, 2004 [77]. With permission.)

PS/CO$_2$ system. Thus, minimization of the interfacial surface energy between talc and polymer matrix would therefore necessitate the depletion of the plasticizing PFA molecules over a certain layer thickness surrounding the talc particle. Such segregation was observed for other multiple-component systems, as for the ternary immiscible poly(methyl methacrylate)/poly-styrene/high-density polyethylene (PMMA/PS/HDPE) polymer blends [68]. The PMMA droplets were encapsulated by PS shells having a maximum thickness such that these composite droplets behave as pure PS droplets, and these droplets were dispersed in the HDPE matrix. This complex morphology was totally driven by the minimization of the overall surface energy.

This migration would therefore increase localized concentration of PFA, as schematically illustrated in Figure 5.23, which was probed with ultra-sound at higher degassing pressures than those associated with nominal values of PFA. For the trials conducted with only 3 wt% of HFC-134a, the impact of the dissolved PFA molecules on the surface energy is large, and yields migration of HFC molecules away from the talc surface in order to form areas of higher PFA concentration. These localized concentrated spots nucleate at higher pressures than the bulk. These spots do not interact for talc content of 1 and 2 wt%, and a plateau in the degassing pressures is observed. However, with 3 wt% of talc, the density of particulates is such that intersection of concentrated spots yields areas with even higher PFA content, which are then associated with much higher degassing pressures.

At 6 wt% of HFC-134a, degassing occurs at a pressure close to that of the critical pressure of the PFA. Variation of the concentration in that region has less impact on the surface energy [67], and migration, even less favorable, is still significant, as can be seen from the slight increase of the degassing pressures with 1 wt% of talc (Figure 5.20).

5.4.4   Impact of Screw Configuration on PFA Dissolution

In twin-screw extrusion, the dissolution of the PFA is performed over a limited number of barrel sections, simultaneously with the cooling of the melt. With the blowing agent dissolved in the polymer, the plasticization effect enables significant temperature reduction, which may unfortunately slow down the diffusion of the PFA molecules in the mixture. Efficient dissolution must then be provided rapidly, and this is feasible through adequate mixing of the components. The screw configuration is then an important factor for these extrusion processes.

The efficiency of four screw configurations, displayed in Figure 5.24, was tested on a co-rotating twin-screw extruder, with different mixtures of HFC-134a/PS (up to 7 wt% of PFA). HFC-134a is particularly known to diffuse very slowly in PS. To make the task more difficult for the various screw functions, the residence time allowed for dissolution was kept to a minimum with the injection of the foaming agent, using a preparative chromatography pump, located in barrel #10. Extrusion was performed at 10 kg/h and 200 rpm. Distributive mixing in screw #1 was obtained through the use of gear and combing mixers, while intensive shear mixing was provided in screw #3 and screw #4 by a series of kneading blocks following the combing mixer. No mixing element was incorporated in screw #2. Reverse elements were located upstream of the injection zone in order to build a melt seal and thus prevent any losses of the blowing agent towards the feed throat. Assessment of the screw efficiency was made through the monitoring of the ultrasonic properties, with an instrumented die equipped with ultrasonic transducers installed at the exit of the extruder.

Results for the ultrasonic velocity, corrected for any fluctuations in pressure or temperature, are displayed in Figure 5.25 as a function of the foaming agent content. Since smaller ultrasonic velocity values indicate enhanced plasticization, configuration #1 yields the best mixing and dissolution, with a continuous decrease of the ultrasonic velocity even for concentrations of HFC-134a as high as 6 wt%. Poor homogeneous mixing of PFA with the PS matrix is observed for all other screw configurations that have led to higher velocity values. In the case of screw configuration #4, leakage at the melt seal was experienced with PFA concentrations in the range of 4 wt%. Without any mixing element in its configuration, screw #2 predictably led to the worst results at low PFA content.

Adequate dissolution and homogeneous distribution of the PFA in the polymer is critical, in order to avoid high-content local spots that would

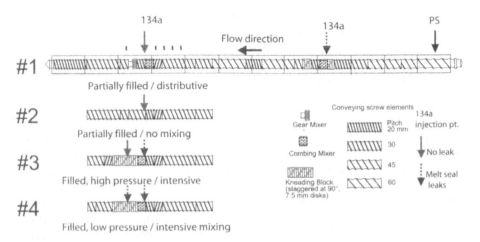

FIGURE 5.24
Configurations of the screws tested for injection of HFC-134a showing locations of injection ports and leakage at the melt seal. (From Gendron, R. et al., *Cell. Polym.*, 21, 315, 2002 [53]. With permission.)

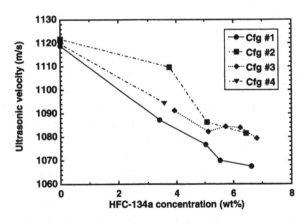

FIGURE 5.25
Plasticization as detected by ultrasonic velocity for the different screw configurations as a function of HFC-134a concentration. Velocities are corrected for pressure variation. (From Gendron, R. et al., *Cell. Polym.*, 21, 315, 2002 [53]. With permission.)

induce premature nucleation and eventually lead to blowholes and a poor foam structure. In the absence of any abnormal ultrasonic signals, it was not possible to assess the existence of a second phase in the melt associated with any undissolved HFC-134a fraction. With accurate weight monitoring of the HFC cylinder over time, implying that the concentrations reported represent adequately the mass of HFC-134a being injected and conveyed by the flowing polymer up to the die exit, the existence of concentration gradients in the mixture were suspected. An irregular degree of plasticization throughout the bulk state is therefore the source of the higher than expected value of ultrasonic velocity associated with some inefficient screw configurations.

5.4.5   Chemical Blowing Agent Decomposition (Off-Line)

Simultaneous reactions occur during the injection molding process of crosslinked low-density foams, including the thermal decomposition of the chemical blowing agent (CBA), the crosslinking induced by peroxide, and the interactions between the reactive components. While the kinetics of the reactions can be studied through DSC measurements [69,70], the ultrasonic static technique mimics more adequately the conditions that prevail during the molding process. First, CBA decomposition and crosslinking are induced as the temperature of the melt rises above some threshold temperatures, with a high pressure maintained in order to prevent phase separation. After an adequate curing period, pressure is finally reduced (this stage mimics the mold opening in the injection process) to induce nucleation, bubble growth, and expansion of the foam. While the small sizes used for the DSC scan can encounter homogeneity problems, much larger samples of approximately 5 g are used for ultrasonic characterization. Sound velocity and attenuation are continuously monitored, which allows for the determination of the onset of CBA decomposition, decomposition rate, diffusion, homogenization, and degassing conditions [71]. Such experiments demonstrate the complexity of CBA decomposition and of gas molecule diffusion in the polymer matrix.

The ultrasonic technique was tested on foamable compounds prepared from poly(ethylene-co-octene) resin (Engage™ from DuPont Dow Elastomers), having different dicumyl peroxide contents. The content of the CBA (azodicarbonamide, Celogen™, AZ130, from Uniroyal Chemical) was kept constant at 5.2 wt%. The CBA decomposition temperature (205 to 215°C) was decreased through the addition of zinc oxide and zinc stearate, in a temperature range close to that of the decomposition of the peroxide and radical formation (140°C). Talc (4.3 wt%) was also added for nucleation purposes. The characterized was performed through a temperature sweep conducted at 2°C/min up to 160°C, at a constant pressure high enough (30 MPa) to maintain the gases in solution formed through the decomposition of the CBA.

First, the ultrasonic technique was not sensitive enough to probe the effects of both exothermic decomposition of the peroxide, which occurs at a temperature close to 142°C [73], and the crosslinking reaction that follows, despite the fact that ultrasound can usually probe crosslinking, as in the case of highly reticulated thermosets [74].

However, with the addition of CBA, specific volume, attenuation, and sound velocity show some specific features, as illustrated in Figure 5.26. These effects can be directly related to the decomposition of the CBA, the diffusion of the gas molecules, and the homogenization of the material. For instance, the abrupt increase of the specific volume is the result of the swelling

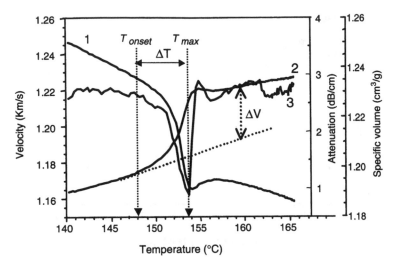

FIGURE 5.26
Variation of velocity (1), specific volume (2), and attenuation (3) when chemical blowing agent (CBA) decomposition occurs. CBA = 5.2 wt%; Peroxide = 0.87 wt%; P = 30 MPa; heating rate = 2°C/min. (From Tatibouët, J. et al., *Polym. Test.*, 23, 125, 2003 [71]. With permission from Elsevier.)

of the polymer chains (and not the cell expansion), as observed for many other gas–polymer systems [5], since the gas generated is kept dissolved in the polymer matrix due to the high pressures maintained (30 MPa).

Typical temperatures are defined in Figure 5.26: T_{onset} is related to the temperature where specific volume, attenuation, and velocity diverge from their linear dependence with temperature; T_{onset} marks the beginning of the CBA decomposition; T_{max}, which coincides with the minimum of the attenuation peak, corresponds to the end of the CBA decomposition process, which also induces the return to linearity for specific volume and velocity. With the assumption that the decomposition rate is constant, $\Delta T = T_{max} - T_{onset}$ is related to the diffusion rate of gas molecules. A small ΔT corresponds to fast dissolution and homogenization. As displayed in Figure 5.27(a), the peroxide content has a strong impact on T_{onset}, this latter being reduced to a lower temperature as the peroxide content is increased, until a plateau value of 142°C is met at a higher peroxide content. This could be explained by the exothermic feature of the crosslinking reaction [75], which has been observed to start roughly at 142°C [73]. As displayed in Figure 5.27(b), ΔT, which accounts for the time required for the gas molecules to diffuse, increases linearly with the level of crosslinking controlled by the peroxide content. This is in agreement with the reported data of diffusivity of small hydrocarbons in irradiation crosslinked low-density polyethylene (LDPE) [76] (these data are also displayed in Figure 1.3).

The mechanisms involved as the compound reaches T_{onset} may explain the sharp decrease in attenuation and velocity together with material swelling (Figure 5.26): CBA decomposition generates residual solids and gas

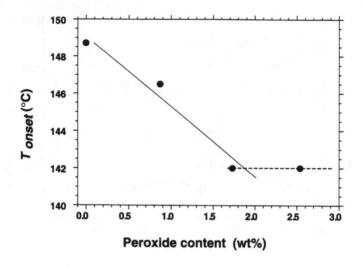

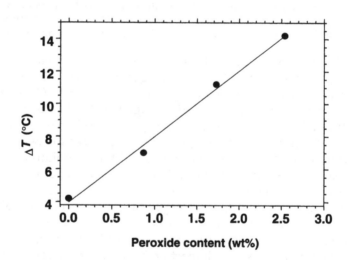

FIGURE 5.27

Evolution of (a) T_{onset} (b) ΔT as functions of peroxide content. (From Tatibouët, J. et al., *Polym. Test.*, 23, 125, 2003 [71]. With permission from Elsevier.)

molecules, inducing a higher concentration of gas molecules around these residual particles. The polymer is locally plasticized and a low viscosity zone temporarily exists around the solid particles and the talc particles located nearby. Subsequently, the diffusion of gas molecules into the polymer occurs, governed by concentration gradients, and the local low-viscosity zone disappears with the material becoming homogeneous, leading to the associated values for ultrasonic propagation parameters and specific volume. These mechanisms are schematically illustrated in Figure 5.28.

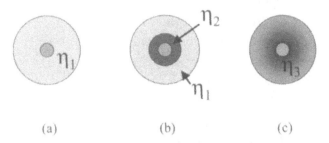

(a)   (b)   (c)

FIGURE 5.28
Schematic description of chemical blowing agent decomposition and gas molecules diffusion, with increasing temperature and time: (a) $T < T_{onset}$. (b) $T = T_{onset}$; low viscosity zone around residual particles is due to enhanced plasticization; attenuation decreases. (c) $T > T_{onset}$; gas molecule diffusion and homogeneization. (From Tatibouët, J. et al., *Polym. Test.*, 23, 125, 2003 [71]. With permission from Elsevier.)

5.5 Conclusions — Future Perspectives?

Numerous characterization techniques used to investigate foam extrusion have been presented in this chapter. Each critical step of the foam extrusion process where the interaction between the polymer and the blowing agent plays the leading role can be studied using either off-line or in-line methods, but the latter should be preferred, because they include complete processing parameters and give a better approach for a global understanding. Among these techniques, the ultrasonic characterization technique permits investigation of each key step of the process, in addition to being able to be used either off- or in-line. Being nondestructive and noninvasive, this easy-to-use technique can be considered not only as an investigating tool but also as an industrial monitoring technique.

Of course, further work has to be done to reach a better understanding of the foam extrusion process. If predictive models exist for foaming agent solubility and plasticization, a more global model for bubble nucleation needs to be developed. At a molecular level, understanding the role of nucleating agents and their interaction with the mixture of polymer and foaming agent will be a great step toward full control of the process.

Abbreviations

CBA:	Chemical blowing agent
DSC:	Differential scanning calorimetry
FFT:	Fast Fourier transform

HCFCs: Hydrochlorofluorocarbons
HFCs: Hydrofluorocarbons
LDPE: Low-density polyethylene
MFI: Melt flow index
NIR: Near-infrared
PCR: Principal component regression
PFA: Physical foaming agent
PLS: Partial least squares
PMMA: Poly(methyl methacrylate)
PS: Polystyrene

References

1. Wong, B., Zhang, Z., and Handa, Y.P., High-precision gravimetric technique for determining the solubility and diffusivity of gases in polymers, *J. Polym. Sci. Part B: Polym. Phys.*, 36, 2025, 1998.
2. Gorski, R.A., Ramsey, R.B., and Dishart, T., Physical properties of blowing agent polymer systems. I. Solubility of fluorocarbon blowing agents in thermoplastic resins, *J. Cell. Plast.*, 22, 21, 1986.
3. Sato, Y. et al., Solubility of hydrofluorocarbon (HFC-134a, HFC-152a) and hydrochlorofluorocarbon (HCFC-142b) blowing agents in polystyrene, *Polym. Eng. Sci.*, 40, 1369, 2000.
4. Daigneault, L.E. et al., Solubility of blowing agents HCFC-142b, HFC-134a, HFC-125 and isopropanol in polystyrene, *J. Cell. Plast.*, 34, 219, 1998.
5. Hilic, S. et al., Simultaneous measurement of the solubility of nitrogen and carbon dioxide in polystyrene and of the associated polymer swelling, *J. Polym. Sci. Part B: Polym. Phys.*, 39, 2063, 2001.
6. Chen, L., Sheth, H., and Kim, R., Gas absorption with filled polymer systems, *Polym. Eng. Sci.*, 41, 990, 2001.
7. Thomas, Y. et al., Monitoring of composition and bubble formation in PS/HCFC mixtures, *J. Cell. Plast.*, 33, 516, 1997.
8. Nagata, T., Tanigaki, M., and Ohshima, M., On-line NIR sensing of CO_2 concentration for polymer extrusion foaming process, *Polym. Eng. Sci.*, 40, 1843, 2000.
9. Zhang, Q., In-line measurement of solubility of physical blowing agents in thermoplastic melts as related to extrusion foaming, Ph.D. thesis, New Jersey Institute of Technology, 2000.
10. Gendron, R. and Daigneault, L.E., Rheology of thermoplastic foam extrusion process, in *Foam Extrusion: Principles and Practice*, Lee, S.-T., Ed., Technomic, Lancaster, PA, 2000, chap. 3.
11. Chow, T.S., Molecular interpretation of the glass transition temperature of polymer-diluent systems, *Macromolecules*, 13, 362, 1980.
12. Zhang, Z. and Handa, Y.P., An *in situ* study of plasticization of polymers by high-pressure gases, *J. Polym. Sci. Part B: Polym. Phys.*, 36, 977, 1998.

13. Hwang, Y.D. and Cha, S.W., The relationship between gas absorption and the glass transition temperature in a batch microcellular foaming process, *Polym. Test.*, 21, 269, 2002.

14. Gerhardt, L.J., Manke, C.W., and Gulari, E., Rheology of poly-dimethylsiloxane swollen with supercritical carbon dioxide, *J. Polym. Sci. Part B: Polym. Phys.*, 35, 523, 1997.

15. Kwag, C., Manke, C.W., and Gulari, E., Rheology of molten polystyrene with dissolved supercritical and near-critical gases, *J. Polym. Sci. Part B: Polym. Phys.*, 37, 2771, 1999.

16. Royer, J.R. et al., High-pressure rheology of polystyrene melts plasticized with CO_2: experimental measurement and predictive scaling relationships, *J. Polym. Sci. Part B: Polym. Phys.*, 38, 3168, 2000.

17. Xue, A. and Tzoganakis, C., Measurement of entrance pressure drop of poly-styrene/supercritical CO_2 solutions, in *Proc. SPE Annu. Mtg. ANTEC*, Dallas, TX, 2001, 1022.

18. Areerat, S., Nagata, T., and Ohshima, M., Measurement and prediction of LDPE/CO_2 solution viscosity, *Polym. Eng. Sci.*, 42, 2234, 2002.

19. Yoo, H.J. and Han, C.D., Studies on structural foam processing. III. Bubble dynamics in foam extrusion through a converging die, *Polym. Eng. Sci.*, 21, 69, 1981.

20. Han, C.D. and Yoo, H.J., Studies on structural foam processing. IV. Bubble growth during mold filling, *Polym. Eng. Sci.*, 21(9), 518, 1981.

21. Taki, K. et al., Visual observations of batch and continuous foaming processes, *J. Cell. Plast.*, 39, 155, 2003.

22. Han, J.H. and Han, C.D., A study of bubble nucleation in a mixture of molten polymer and volatile liquid in a shear flow field, *Polym. Eng. Sci.*, 28, 1616, 1988.

23. Han, J.H. and Han, C.D., Bubble nucleation in polymeric liquids. I. Bubble nucleation in concentrated polymer solutions, *J. Polym. Sci. Part B: Polym. Phys.*, 28, 711, 1990.

24. Cloetens, P. et al., Hard X-ray phase imaging using simple propagation of a coherent synchrotron radiation beam, *J. Phys. D: Appl. Phys.*, 32, A145, 1999.

25. Ryan, A.J., Using time resolved X-ray diffraction at synchrotron radiation sources to study structure development in polymer processing, *Cell. Polym.*, 14, 444, 1995.

26. Szayna, M., Zedler, L., and Voelkel, R., NMR microscopy and image processing for the characterization of flexible polyurethane foam, *Angew. Chem. Int. Ed.*, 38, 2551, 1999.

27. Nakatani, A.I. et al., Shear stabilization of critical fluctuations in bulk polymer blends studied by small angle neutron scattering, *J. Chem. Phys.*, 93, 795, 1990.

28. Rayleigh, J.W.S., *Theory of Sound*, Dover, New York, 414–431, 1945.

29. Waterman P.C. and Truell, R., Multiple scattering of waves, *J. Math. Phys.*, 2, 512, 1962.

30. Lloyd, P. and Berry, M.V., Wave propagation through an assembly of spheres, *Proc. Phys. Soc. London*, 91, 678, 1967.

31. Ma, Y., Varadan, V.K., and Varadan V.V., Comments on ultrasonic propagation in suspensions, *J. Acoust. Soc. Am.*, 87, 2779, 1990.

32. Allegra, J.R. and Hawley, S.A., Attenuation of sound in suspensions and emulsions: theory and experiments, *J. Acoust. Soc. Am.*, 51, 1545, 1972.

33. Tebbutt, J.S. and Challis, R.E., Ultrasonic wave propagation in colloidal suspensions and emulsions: a comparison of four models, *Ultrasonics*, 34, 363, 1996.

34. Piché, L. et al., U.S. Patent 4,754,645, 1988.
35. Sahnoune, A. and Piché, L., Glass transition and ultrasonic relaxation in polystyrene, *Proc. MRS Fall Mtg.*, Boston, 1996.
36. Sahnoune, A., Massines, F., and Piché, L., Ultrasonic measurement of relaxation behavior in polystyrene, *J. Polym. Sci. Part B: Polym. Phys.*, 34, 341, 1996.
37. Piché, L., Ultrasonic velocity measurement for the determination of density in polyethylene, *Polym. Eng. Sci.*, 24, 1354, 1984.
38. Tatibouët, J. and Piché, L., Ultrasonic investigation of semi-crystalline polymers: study of poly(ethylene terephthalate), *Polymer*, 32, 3147, 1991.
39. Piché, L. et al., U.S. Patent 5,443,112, 1995.
40. Gendron, R. et al., Ultrasonic behavior of polymer blends, *Polym. Eng. Sci.*, 35, 79, 1995.
41. Wang, H. et al., On-line ultrasonic monitoring of the injection molding process, *Polym. Eng. Sci.*, 37, 363, 1997.
42. Brown, E.C. et al., An ultrasound virtual instrument approach for monitoring of polymer melt variables, *Proc. SPE Annu. Mtg. ANTEC*, 44, 335, 1998.
43. Shen, J. et al., Ultrasonic melt temperature measurement during extrusion, *Proc. SPE Annu. Mtg. ANTEC*, 44, 2076, 1998.
44. Piché, L. et al., Ultrasonic probe for on-line process monitoring, *Plast. Eng.*, 55(10), 39, 1999.
45. Tatibouët, J., Hamel, A., and Piché, L., Ultrasound as monitoring tool for polymer degradation and reactive extrusion, *Proc. 15ᵗʰAnnu. Mtg. Polymer Processing Society*, Den Bosch, Netherlands, 1999.
46. Tatibouët, J. and Huneault, M.A., In-line ultrasonic monitoring of filler dispersion during extrusion, *Intern. Polym. Proc.*, 17, 49, 2002.
47. Sahnoune, A. et al., Ultrasonic monitoring of foaming in polymers, *Proc. SPE Annu. Mtg. ANTEC*, 2259, 1997.
48. Tatibouët, J. and Gendron, R., A study of strain-induced nucleation in thermoplastic foam extrusion, *J. Cell. Plast.*, 40, 27, 2004.
49. Yang, H.-H. and Han, C.D., The effect of nucleating agents on the foam extrusion characteristics, *J. Appl. Polym. Sci.*, 29, 4465, 1984.
50. Tatibouët, J. et al., Effect of different nucleating agents on the degassing conditions as measured by ultrasonic sensors, *J. Cell. Plast.*, 38, 203, 2002.
51. Lee, S.-T., A fundamental study of thermoplastic foam extrusion with physical foaming agents, in *Polymeric Foams Science and Technology*, Khemani, K.C., Ed., American Chemical Society, Washington, D.C., 1997, chap. 13.
52. Sahnoune, A. et al., Application of ultrasonic sensors in the study of physical foaming agents for foam extrusion, *J. Cell. Plast.*, 37, 429, 2001.
53. Gendron, R. et al., Foam extrusion of polystyrene blown with HFC-134a, *Cell. Polym.*, 21, 315, 2002.
54. Graessley, W.W., The entanglement concept in polymer rheology, *Adv. Polym. Sci.*, 16, 60, 1974.
55. Goel, S.K. and Beckman, E.J., Generation of microcellular polymeric foams using supercritical carbon dioxide-1: Effect of pressure and temperature on nucleation, *Polym. Eng. Sci.*, 34, 1137, 1994.
56. Graessley, W.W., Glasscock, S.D., and Crawley, R.L., Die swell in molten polymers, *Trans. Soc. Rheol.*, 14, 519, 1970.
57. Rohn, C.L., *Analytical Polymer Rheology*, Hanser Publishers, New York, 1995, chap. 8.

58. Kim, S. and Dealey, J.M., Gross melt fracture of polyethylene. I: A criterion based on tensile stress, *Polym. Eng. Sci.*, 42, 482, 2002.
59. Throne, J.L., *Thermoplastic Foams*, Sherwood Publishers, Hinckley, OH, 1996.
60. Colton, J.S. and Suh, N.P., The nucleation of microcellular thermoplastic foam with additives. Part I: Theoretical considerations, *Polym. Eng. Sci.*, 27, 485, 1987; Colton, J.S. and Suh, N.P., The nucleation of microcellular thermoplastic foam with additives. Part II: Experimental results and discussion, *Polym. Eng. Sci.*, 27, 493, 1987.
61. Chen, L. et al., Effect of filler size on cell nucleation during foaming process, *J. Cell. Plast.*, 38, 139, 2002.
62. Areerat, S. et al., Solubility of carbon dioxide in polyethylene/titanium dioxide composite under high pressure and temperature, *J. Appl. Polym. Sci.*, 86, 282, 2002.
63. Ramesh, N.S., Rasmussen, D.H., and Campbell, G.A., The heterogeneous nucleation of microcellular foams assisted by the survival of microvoids in polymer containing low glass transition temperature particles. Part I: Mathematical modeling and numerical simulation, *Polym. Eng. Sci.*, 34, 1685, 1994.
64. Lee, S.-T., Shear effects on thermoplastic foam nucleation, *Polym. Eng. Sci.*, 33, 418, 1993.
65. Jones, R.A.L. and Richards, R.W., *Polymers at Surfaces and Interfaces*, Cambridge University Press, Cambridge, U.K., 1999, chap. 5.
66. Yildirim, I., Surface free energy characterization of powders, Ph.D. thesis, Virginia Polytechnic Institute and State University, 2001.
67. Jaeger, P.T., Eggers, R., and Baumgartl, R.W., Interfacial properties of high viscous liquid in a supercritical carbon dioxide atmosphere, *J. Supercrit. Fluids*, 24, 203, 2002.
68. Reignier, J. and Favis, B.D., On the presence of a critical shell volume fraction leading to pseudo-pure droplet behavior in composite droplet polymer blends, *Polymer*, 44, 5061, 2003.
69. Prasad, A. and Shanker, M.A., A quantitative analysis of chemical blowing agent by DSC, *Proc. SPE Annu. Mtg. ANTEC*, 1860, 1998.
70. Cassel, B. and Sichina, W.J., Characterization of foams by thermal analysis, *Thermal Trends*, 6, 10, 1999.
71. Tatibouët, J., Gendron, R., and Haïder, L., Ultrasonic characterization performed during chemical foaming of cross-linked polyolefins, *Polym. Test.*, 23, 125, 2003.
72. Datta, D. et al., An ultrasonic technique to monitor the blowing process in sponge rubbers, *Polym. Test.*, 21, 209, 2002.
73. Vachon, C. and Gendron, R., Effect of viscosity on low density foaming of poly(ethylene-co-octene) resins, *J. Cell. Plast.*, 39, 71, 2003.
74. Legros, N., Jen, C.-K., and Ihara, I., Ultrasonic evaluation and application of oriented polymer rods, *Ultrasonics*, 37, 291, 1999.
75. Moulinié, P. and Woelfle, C., Investigation of the reaction kinetics within expandable mixtures used for preparing injection-molded polyolefin foams, *Proc. SPE Annu. Mtg. ANTEC*, 2031, 1999.
76. Pauly, S., Permeability and diffusion data, in *Polymer Handbook*, 4th ed., Brandrup, J., Immergut, E.H., and Grulke, E.A., Eds., John Wiley & Sons, New York, 1999, chap. 6.
77. Tatibouët, J. and Gendron, R., Nucleation in foams as assessed by in-line ultrasonic measurements, *Proc. SPE Annu. Mtg. ANTEC*, 2552, 2004.

6

The Relationship between Morphology and Mechanical Properties in Thermoplastic Foams

Martin N. Bureau

CONTENTS

6.1 Introduction

Foams, as engineering materials, are now used in all industrial sectors and represent an extraordinary class of materials. They are intended for packaging, thermal and electrical insulation, buoyancy, and structural applications such as decks, road pavements, sandwich panels, and so on. This extraordinary character comes from their diverse functionalities — stiffness, strength, impact resistance, dielectric and thermal resistance, and permeability, among others — which can be customized to obtain properties ranging beyond the limits of all other classes of engineering materials. A great challenge in engineering foams is to address parameters such as specific weight and cost while answering functional requirements.

Polymeric foams are generally characterized in terms of mechanical performance by means of different conventional mechanical testing methods. The most popular of these methods are summarized in Table 6.1. While these mechanical characterization techniques can help to produce a foam that will perform according to certain functional requirements, additional work still needs to be done in order to understand the origin of their performance. To achieve this, the relationship between the morphology of engineering foams and their functional behavior needs to be defined. Until now, the relationship between morphology and performance has been discussed according to the approach proposed by Gibson and Ashby [1], which relates foams' mechanical properties to their density (see Section 6.2). However, this approach, based exclusively on the foam density, does not take into account the detailed morphological characteristics of foams, such as their cell size distribution and anisotropy. In recent years, some attempts at considering detailed morphological foam characteristics in mechanical behavior analysis have been reported; these are reviewed below (see Section 6.3). However, this review is not intended to provide an exhaustive survey of foam properties, nor a comparison of the performance of different foams or foamed structures. One source for such a treatment is the excellent textbook of Gibson and Ashby [1], which provides a very wide survey of foam structures and properties. A microstructural description of foam materials that takes into account their detailed morphological characteristics and that can be employed as a single microstructural parameter to report mechanical properties will be provided

TABLE 6.1

Summary of the Most Popular Mechanical Testing Methods Used for Foams

Testing Method Designation	Test Method Description	Ref.
Compression	Constant section (square or circular) specimen compressed at constant nominal strain rate until specimen failure or until the deformation reaches the nominal void fraction.	2
Flexural	Constant rectangular section beam specimen in three- or four-point bending until specimen failure or load plateau.	3
Falling dart impact	Clamped plate specimen with a circular rig impacted by a falling dart equipped with a load cell and a displacement transducer (impact should generate failure of the specimen).	4
Dynamic shock cushioning (flat sheet impact)	Constant section (square or circular) specimen impacted by a plateau at high speed until specimen failure or the deformation reaches the nominal void fraction (test may be repeated to obtain maximum compression or densification; see Section 6.2.1).	5

in Section 6.4. Mechanical results obtained from different polystyrene (PS) and polyolefin foams are analyzed using the proposed approach, considering only cell size distribution at first, and then cell size distribution and anisotropy in Section 6.5. Finally, the proposed approach is summarized and presented as an integrated useful tool in the context of foam production and design.

6.2 Classical Mechanical Approach

6.2.1 Standard Mechanical Testing

The most popular mechanical testing method used for foams is the compression test. This testing method [2] consists of compressing a foam specimen and recording the load reaction as the specimen reduces in thickness as a result of the imposed displacement. The compressive stress–strain curve is obtained by dividing the load at any moment during the test by the nominal area and the imposed reduction in thickness by the initial thickness of the specimen. An illustration of this compressive stress–strain curve is shown in Figure 6.1. This compressive stress–strain curve typically shows three stages. The first stage is characterized by a linear elastic or Hookean behavior where the stress is proportional to the strain, following Hooke's law. The compressive modulus of elasticity E^* is obtained from the slope of this initial linear portion of the compressive stress–strain curve. The compressive strength σ_y^* is obtained from the compressive stress at the yield point if a yield point occurs before 10% deformation or, in the absence of such a behavior, from the compressive stress at 10% deformation. The compressive yield point is defined as the first point on the compressive stress–strain curve at which an increase in strain occurs without an increase in stress. Starting from the yield point, the second stage of the compressive stress–strain curve is characterized by an approximately constant compressive stress rate (plateau) during which the foam undergoes plastic deformation, either by elastic buckling, plastic yielding, or brittle crushing, depending on the foam behavior. This constant compressive stress rate stage extends over a certain amount of compressive deformation and ends at the onset of densification, which defines the third and last stage of the compressive stress–strain curve. The densification stage is a progressive one, during which the foam cells collapse completely, resulting in a rapid increase of the compressive stress as a result of the densification of the foam. The limit of this densification stage in terms of deformation is set by the amount of material in the foam, i.e., its density. As a first approximation, this deformation at densification ε_d^* may be obtained from the void fraction in the foam, given by $1 - \rho_r$ (where ρ_r is the relative foam density). As the foam is collapsing, the compressive stress–strain curve progressively tends towards the bulk modulus of the material in the foam.

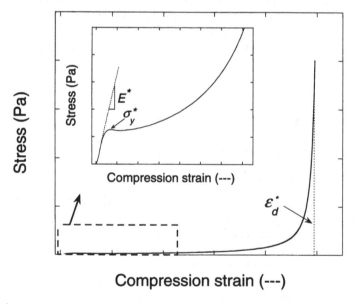

FIGURE 6.1
Typical compression stress–strain curve of a low-density elasto-plastic foam, with compressive modulus of elasticity E^*, compressive strength σ_y^*, and strain at densification ε_d^*, graphically represented.

Other testing methods have also been used to characterize the mechanical performance of foams. Flexural tests [3], either in three-point or four-point bending configurations, can be performed. The advantage of the latter is that it permits the researcher to obtain the flexural modulus and the flexural strength, which are of prime interest in the production of foam cores in sandwich panels, for example. It also allows one to obtain the mechanical performance of foams under a positive tensile stress, which is very difficult to obtain for polymeric foams in tensile testing due to stress concentration effects at the grips used to hold the tensile specimens. Variations of impact testing, either with a falling dart [4] or a flat sheet impact (dynamic shock cushioning) [5] are also of interest for quantifying the foam's ability to resist high velocity puncture by an object or to absorb dynamic compressive shocks and retain its properties.

6.2.2 The Classical Approach

As mentioned previously, the standard reference on the morphology and performance relationship is the work of Gibson and Ashby [1], which relates foam properties to foam density. This approach utilizes the *optimization theory for materials selection in design* [6] as a basis for understanding and comparing foam properties. It consists of defining an objective function, which establishes the relationship between shape, density, and required properties for a specific mechanical application. The derivative of this objective function leads to an optimal design. An illustration of this approach is shown in Figure 6.2. As shown by guidelines for constant E/ρ, $E^{1/2}/\rho$, and $E^{1/3}/\rho$ in Figure 6.2, very different classes of materials present a similar specific compressive modulus of elasticity. For example, the regions in Figure 6.2 corresponding to polymeric foams, engineering polymers, porous ceramics, engineering alloys, and engineering ceramics are all intercepted by the same guideline corresponding to $E^{1/3}/\rho = C_1$, indicating that plates made of these materials have a similar specific rigidity. The regions in Figure 6.2 corresponding to polymeric foams and engineering polymers are also intercepted by the same guideline of $E^{1/2}/\rho = C_1$, indicating that beams made of these materials have a similar specific rigidity. Although it permits us to compare very different classes of materials, this approach is highly macroscopic and does not consider the particularities of each class of materials with respect to their morphology and architecture. At the basis of the success of this approach is the fact that materials are defined by their density and a given property, which define a region in the materials selection chart that in most cases does not overlap from one class of materials to another.

Gibson and Ashby [1] have developed this approach further to express foam properties as a function of density. The latter considers a simple open-cell foam modeled as a cubic array of rigid members of a given length and square cross-section, for which the compressive modulus of elasticity and compressive strength is given by the following:

$$\frac{E^*}{E_o} = C_1 \cdot \left(\frac{\rho^*}{\rho_o}\right)^2 \qquad\qquad (6.1)$$

$$\frac{\sigma_y^*}{\sigma_{y,o}} = C_2 \cdot \left(\frac{\rho^*}{\rho_o}\right) \qquad\qquad (6.2)$$

where superscript * and subscript o designate the foam and base material, respectively; σ_y represents the compressive strength; and C_1 and C_2 represent constants of proportionality. This model can also be extended to closed-cell foams by considering the presence of cell walls between the rigid members and the contribution of gas pressure within foam cells, for which the compressive modulus of elasticity and compressive strength are given by:

$$\frac{E^*}{E_o} = C_3 \cdot \phi^2 \left(\frac{\rho^*}{\rho_o}\right)^2 + C_4 \cdot (1-\phi)\frac{\rho^*}{\rho_o} + \frac{p_i\left(1-2\upsilon^*\right)}{E_o\left(1-\rho^*/\rho_o\right)} \qquad\qquad (6.3)$$

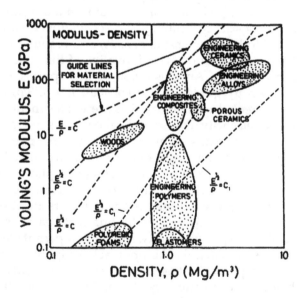

FIGURE 6.2
Materials selection chart of the compressive modulus of elasticity E, plotted against the density ρ. The contour lines define properties for specific classes of materials. The guidelines of constant E/ρ, $E^{1/2}/\rho$, and $E^{1/3}/\rho$ allow the selection of materials for minimum weight deflection-limited design, given by C and C_1, in the case of columns or ties, beams, and plates, respectively. (From Ashby, M.F., *Materials Selection in Mechanical Design*, Butterworth-Heinmann, Oxford, U.K., 1992. With permission from Elsevier.)

$$\frac{\sigma_y^*}{\sigma_{y,o}} = C_5 \cdot \left(\phi \frac{\rho^*}{\rho_o} \right)^{3/2} + C_6 \cdot (1-\phi) \frac{\rho^*}{\rho_o} + \frac{p_i - p_{at}}{\sigma_{y,o}} \tag{6.4}$$

where ϕ is the fraction of solid in the cell edges; p_i and p_{at} are the gas pressure in the foam cells and the atmospheric pressure; υ^* is the Poisson's ratio of the foam; and C_3, C_4, C_5, and C_6 represent constants of proportionality.

Using Equations 6.3 and 6.4, the relative compressive modulus of elasticity and the relative compressive strength can be expressed as a function of the relative density. An illustration of this is given in Figure 6.3. This figure shows that the properties of different foams extending over two decades of relative densities (from 0.01 to 1) can be scaled up using this approach. A closer look at the data in Figure 6.3 reveals, however, that at a given relative density the properties can vary up to one order of magnitude. Given this scatter in the data and the logarithm scales used in this figure, the materials selection approach cannot account for nor predict variations in properties within specific classes of foams, or materials in general, or specific ranges of densities, as it only defines a material in terms of density and does not take into account the microstructure of the material considered. Thus, polymeric foams with different cell size distribution but similar density are expected, using this approach, to have the same compressive modulus of elasticity and compressive strength, but this does not represent the properties of such foams in reality, as will be shown in Section 6.5. A refinement of the description of the mechanical behavior of foams, based on the characteristics of their morphology, is thus necessary. Attempts to address this question have been reported in recent years and are reviewed briefly in the next section.

6.3 Literature Review

Other than the textbook of Gibson and Ashby [1], very few publications have done more in terms of the mechanical behavior study of engineering polymeric foams than report some of their basic mechanical properties, for example, the compression set or the elastic modulus in compression. From 1997 to the present, the number of scientific journal papers on the relationship between foam morphology and mechanical behavior has been less than 50, and this figure includes microcellular foam studies. The most interesting studies among them, in the opinion of the author, are briefly reviewed herein.

6.3.1 Microcellular Foams

Microcellular foams (MCFs) are foams with an average cell size on the order of 10 µm and very high cell nucleation density (e.g., $\approx 10^9$ cells/cm³), both of

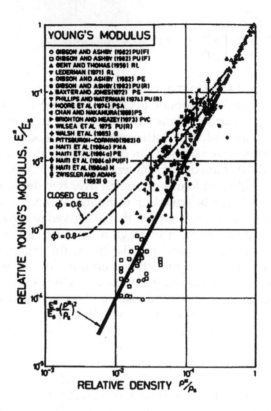

(a)

FIGURE 6.3

Data of the relative compressive modulus of elasticity and the relative compressive strength of foams plotted against relative density. The solid lines represent the theoretical expression of Equations 6.1–6.4. (From Gibson, L.J. and Ashby, M.F., *Cellular Solids: Structure and Properties*, 2nd ed., Cambridge University Press, Cambridge, U.K., 1997. Reprinted with the permission of Cambridge University Press.)

which result in a significant density reduction, typically on the order of 30%. The process for producing MCFs was developed by Martini, Suh, and Waldman [7], using thermodynamic instability to cause bubble nucleation and growth in the solid state. This microcellular foaming process was performed by Kumar and Suh using a two-stage approach [8] that involves gas saturation of a polymer at elevated pressure and room temperature, followed by rapid pressure drop to atmospheric pressure, causing gas supersaturation, and heating to a temperature near the glass transition temperature to soften the polymer. Since microcellular foams are claimed to present mechanical properties not significantly different from those of the unfoamed matrix, due to their cellular structure, which is smaller than an inherently critical flaw size, several studies on their mechanical behavior have been reported in the last ten years. Some of them focused exclusively on their basic mechanical properties, such as their modulus of elasticity or their strength, which are

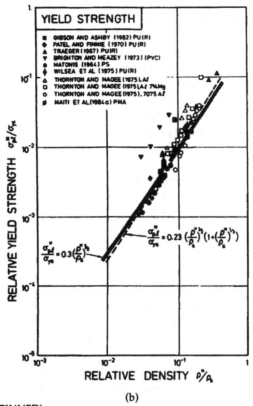

FIGURE 6.3 (CONTINUED)

proportional to their density [9–19], in agreement with Gibson and Ashby (see Section 6.2.2), and others concluded that certain properties, such as impact strength, are improved in MCFs in comparison with the unfoamed polymer [20–22].

Among the group of studies concluding that properties are improved in MCFs, one study on polycarbonate (PC) MCFs [20] reports that an increase in the notched Charpy impact strength, both in terms of load and energy to fracture, is obtained in an MCF with an average cell size of 40 µm and a density reduction of more than 20%, with respect to either unfoamed PC, or to PC-MCF with an average cell size of a 20 µm or less and a density reduction of less than 5%. However, the results of this study illustrated the well-known notch sensitivity of PC and focused only on results obtained from Charpy specimens with very sharp notches produced using a high-speed cutting tool. No precracking procedure was used prior to impact testing, contrary to recommendations for notch sensitive materials in fracture testing methods [23,24]. It is most likely that the notching procedure produced local heating and melting at the notch root, and that the latter produced different notch tip geometries in the presence of voids of different sizes. Thus, comparison of the Charpy and Izod impact responses of such materials is quite uncertain and should be kept reserved for quality control and materials specification, as stated in the "Significance and Use" note in the *ASTM Charpy or Izod Standards* [25,26]: "The fact that a material shows twice the energy absorption of another under these conditions of test does not indicate that this same relationship will exist under another set of test conditions. The order of toughness may even be reversed under different testing conditions." Definitive conclusions from such impact testing should thus be based on careful specimen preparation, which should include fatigue precracking or at least fresh razorblade cuts at the notch root, as recommended in fracture testing methods [23,24], and notch root geometry characterization.

Other materials were also studied using the same methodology [21,22] and revealed contradictory results from slow and high speed fracture testing in PS, styrene–acrylonitrile copolymer (SAN), and PC, which might be related to the discussed effects. To support the discussion of these results, a study on the impact strength of PC-MCF [13] reported that, for a constant relative density of 0.7, the impact strength from notched Izod testing varied by more than 100% when cell size changed from 7 to 18 μm. The reported optimal impact resistance for very different cell sizes (less than 20 μm vs. more than 40 μm) in the previous study [20] on similar PC-MCFs of similar relative densities is most probably an artifact of the differences in specimen preparation and testing conditions. Another source of the differences between fracture and impact studies originates from the anisotropic properties and characteristics of MCFs, both in terms of cell size and aspect ratio, which are reflected in the impact strength of MCFs [12]. Crystallinity in MCFs of semicrystalline polymers also plays a role and should be considered [27].

It thus appears that the specific morphological characteristics of MCFs have a great influence on their mechanical properties, both at low and high strain rates, and that the cell size, among other microstructural parameters, plays an important role in the determination of these properties.

6.3.2 Regular Low-Density Foams

Some attempts have been made at predicting the properties of cellular materials using continuum mechanics approaches and finite element modeling. One study [28] reported accurate predictions of Young's modulus and shear modulus as a function of void fraction (i.e., density) based on the Mori-Tanaka model, which considers the internal stress in the matrix of a material containing second phase particles, assimilated to voids for foams, and other continuum mechanics models, which consider stress and displacements, in the case of cavities embedded in a matrix, and solve for continuity at interfaces in the materials. Similar results were obtained in a study [29,30] based on the Mori-Tanaka model, taking into account not only the void fraction but also the void aspect ratio. Results of the model showed that good agreement could be obtained between theoretical and experimental Young's modulus and shear modulus as functions of void fraction, and that the theoretic effect of void aspect ratio on both moduli could be obtained. The latter was not verified experimentally. While these numerical studies could be of interest to predict foam properties, none actually considers the foam architecture, i.e., cell size, cell wall dimensions, and anisotropy, probably because very few experimental studies or data have been reported on this subject.

One of these experimental studies based its approach on the mechanical contribution of internal gas pressure depending on cell size to the overall foam properties [31]. In this study of metallocene polyolefin elastomers and polyethylene blends with different cell size distributions but similar densities, higher compressive strength at high deformation were reported for those foams with a lower mean cell size. The results were not analyzed in detail in the latter study but it was proposed that the contribution of internal pressure due to the gas contained within the cells is lower for larger cells due to the higher probability of breaking for the more strained cell walls. In other words, the external pressure required to overcome the internal gas pressure is higher for smaller cells. From Equations 6.3 and 6.4, however, it appears that the contribution of the internal pressure to the overall mechanical properties of foams is significant only when the internal pressure is extremely high. For example, to obtain a value of $E/E_o = 0.1$, the pressure-dependent third term in Equation 6.3, given by $p_i(1 - 2v^*)/E_o(1 - \rho^*/\rho_o)$, the value of the internal gas pressure in the foam cells, p_i, should reach the astronomic number of 200 MPa. It is not considered that this contribution is significant in reality and therefore the terms related to internal pressure could be neglected.

One of the most significant contributions to the understanding of the mechanical behavior of engineering polymeric foams based on the characteristics of their morphology originates from the work of Rodriguez-Perez and coworkers [32–39], who studied the mechanical behavior of crosslinked closed-cell low-density polyethylene (LDPE) foams blended with various amounts of linear low-density polyethylene (LLDPE), high-density polyethylene (HDPE), poly(ethylene vinyl acetate) (EVA), and isoprene–styrene (IS) copolymers. The results reported provide support for the Gibson and Ashby relationship between foam properties and density [32,33,35,37,38]. The modulus of elasticity and strength were shown to be higher at higher densities by different characterization methods, such as dynamic mechanical analysis (DMA), compression testing at low and high strain rates, and high speed indentation testing (falling dart impact). A linear relationship between the indentation stress and the density, and between the impact modulus and the density, was observed in LDPE-based foams [33,35,36]. These properties were also affected by the amount and properties of the second polymer blended into LDPE. The highest strength and modulus were obtained with HDPE added, while the highest toughness was obtained with a rubber-like phase, EVA or IS, added. Crystallinity of the foam was also noted to have an effect on the foam's mechanical properties [32], but it was difficult to discriminate from the effect of blending. The influence of the cell architecture has also been an object of interest for the same group. The average cell size, the average cell wall thickness, and the fraction of mass in the struts of individual cells were measured [32,35,37,38]. It was reported that no obvious relationship could be established between the density and the cell size or the cell wall thickness.

To address this question specifically, a systematic study of the cell size effect on the mechanical properties of LDPE foams was performed [39]. Foams with nominal densities of 24 and 42 kg/m³ and with different cell sizes and cell wall thicknesses for each nominal density were characterized. The cell size varied from 315 to 965 μm and the cell wall thickness was between 1.6 and 7.0 μm. For all foams considered, the compressive modulus of elasticity and strength were the same at a given density, irrespective of their cell size. It was concluded that the cell size had no effect on the mechanical properties of foams, which were fixed by the foam density. However, this conclusion might need to be revised in light of the following. Plotting the cell size and cell wall thickness against the foam density in Figure 6.4(a) shows that these two microstructural parameters followed the same trends as a result of processing. If we make the simple assumption that the number of cell walls per unit length n_L is given by

$$n_L = \frac{1}{\bar{d}} \tag{6.5}$$

where $\bar{d}$ is the mean cell size, then the amount of material per unit length in a foam is given by

$$N_L = t \cdot \frac{1}{\bar{d}} \tag{6.6}$$

where t is the cell wall thickness, and consequently the amount of material per unit surface N_S can be obtained from

$$N_s = \left(t \cdot \frac{1}{\bar{d}} \right)^2 \tag{6.7}$$

Plotting this amount of material per unit surface N_S calculated from the measured cell size and cell wall thickness against the foam density [39] in Figure 6.4(b) reveals that the same amount of material per unit surface was observed for all foams at a given density, as a result of processing. Similar compressive modulus of elasticity and strength were obtained for the foams with similar N_S values. Consequently, the conclusions of this study are incomplete. The statement concerning the lack of effect of cell size on the mechanical properties of foams should be replaced by one stating that foams

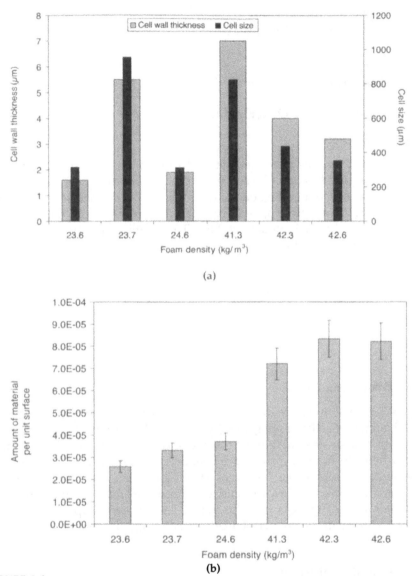

FIGURE 6.4
Cell wall thickness and cell size (a) and calculated amount of material per unit surface (b) plotted against foam density from the microstructural data of PE foams in Rodriguez-Perez, M.A. et al., *Cell. Polym.*, 21, 165, 2002 [39].

with different morphologies could not be prepared and thus no conclusion concerning the effect of cell size or cell wall on the mechanical properties can be drawn. It is clear from this analysis that looking simply at a given microstructural parameter, e.g., the density, is not enough, and that such analysis based on another single microstructural parameter, e.g., the cell size, can lead to incorrect conclusions. Such microstructural analysis must be done, therefore, with great care. The following section presents a way of rationalizing the microstructural characteristics of foam.

6.4 Morphological Description of Foams

6.4.1 Foam Density

As trivial as it may appear, the first and most often used microstructural parameter required to describe a foam is the amount of material it contains. This characteristic, depending on the research field, is defined as the absolute density, the relative density ρ_r, or the void fraction of the foam f. It is generally determined by the Archimedes method. The relative density and the void fraction are defined as follows:

$$\rho_r = \frac{\rho^*}{\rho_o} \tag{6.8}$$

$$f = 1 - \rho_r \tag{6.9}$$

6.4.2 Foam Structure

The next parameter used to describe a foam is the fraction of open and closed cells it contains. The latter are generally measured by comparing the external volume of a foam specimen to the pressurized gas volume differential measured when the foam is put in a hermetic chamber. If both volumes are equivalent, the foam has a 100% closed cell structure. If these volumes are different, their difference is attributed to the open-cell fraction, which permits the penetration of the gas into the foam. The present chapter deals with industrial foams in which a closed-cell morphology predominates.

6.4.3 Cell Size Distribution

Another very important microstructural parameter required to describe a foam is the cell size distribution, from which different statistical parameters can be derived. The cell size distribution is obtained from quantitative observations of foams cells. These cells are observed on the microtomed or fractured surface of the foam. By image analysis, the surface area of individual cells, A_i, is measured and the corresponding cell size, d_i, is calculated from the equivalent diameter of a circle of area A_i ($d_i = 2\sqrt{A_i/\pi}$). To reflect the foam microstructure with a good statistical representation, the number of cells to be measured should be reasonably high, e.g., 200–300. By plotting the cell

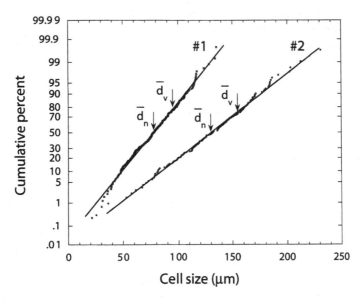

FIGURE 6.5
Cumulative normal distribution of cell sizes for two polystyrene foams: foam #1 with $\overline{d}_n = 77.6\ \mu m$, $\overline{d}_v = 94.5\ \mu m$ ($\overline{d}_v/\overline{d}_n = 1.22$), and $r^2 = 0.996$; foam #2 with $\overline{d}_n = 130.3\ \mu m$, $\overline{d}_v = 155.0\ \mu m$ ($\overline{d}_v/\overline{d}_n = 1.19$), and $r^2 = 0.995$.

size in a histogram, the cell size distribution curve can be obtained. An example of the latter is shown in Figure 6.5. From such a distribution, the number-average diameter $\overline{d}_n$ and the volume-average diameter $\overline{d}_v$ can be calculated using the following:

$$\overline{d}_n = \frac{\sum n_i d_i}{\sum n_i}$$

(6.10)

$$\overline{d}_v = \frac{\sum n_i^4 d_i^4}{\sum n_i^3 d_i^3}$$

(6.11)

where d_i and n_i are a measured diameter i and the number of such diameters measured, respectively.

A cell size distribution that is close to perfectly normal will show very close values of $\overline{d}_n$ and $\overline{d}_v$, while other distributions, a bimodal distribution for example, will show very different values of $\overline{d}_n$ and $\overline{d}_v$. Thus, the ratio of $\overline{d}_v/\overline{d}_n$ represents a useful tool for evaluating cell size dispersity, similar to the molecular weight polydispersity used in polymer chemistry to characterize molecular weight distribution. It is generally believed that a cell size dispersity ratio close to 1 (e.g., below 1.25) indicates a statistically normal distribution or a monodisperse distribution. The distributions in Figure 6.5 can be considered statistically normal. According to the Schwartz–Saltikov statistical method for the correction of measured diameters [40,41], the diameter correction for a monodisperse distribution is given by a constant equal to $4/\pi$ (≈ 1.27). Since the correction factor is a constant for all foams with cell size dispersity ratios close to 1, it will not be considered in the morphological characterization of the foams presented in the next sections.

6.4.4   Foam Anisotropy

Foams produced as extruded panels generally exhibit morphological anisotropy, which affects their mechanical behavior. Depending on the application, it could be necessary to obtain their cell size distribution, average aspect ratio, and orientation in three orthogonal directions, i.e., in the extrusion or machine direction, in the direction normal to the machine direction, and in the direction normal to the thickness of the panels. These three microstructural parameters account for their morphological anisotropy, which will be reflected in their mechanical anisotropy. The principal microstructural parameters are summarized in Table 6.2.

6.4.5   Microstructure Density Parameter

Although these parameters can describe well the characteristics of a given foam, the comparison of different foams, as shown in Section 6.3.2, requires that the density as well as the morphology is taken into account. The basis for comparison used in Section 6.3.2 was defined as the amount of material per unit surface N_S (Equation 6.7). This amount of material was evaluated from the inverse of the average cell size, which corresponds to the number of cells per unit length, and the average cell wall thickness, which corresponds to the quantity of material in a cell. While the average cell size is relatively easy to obtain accurately (see above), the average cell wall thickness requires considerably more care and its quantification is quite problematic. The first problem encountered is that the cell wall thickness is not constant for a given cell; cell faces are generally thinner than cell edges or

TABLE 6.2

Useful Microstructural Parameters to Describe Foams

Foam Characteristics	Microstructural Parameters
Amount of material	Density, relative density, void fraction
Foam structure	Fraction of open and closed cells
Cell size distribution	Number-average and volume-average diameters, cell size dispersity
Anisotropy	Cell size in orthogonal directions, cell aspect ratio, and cell orientation*

* Orientation of the maximum cell diameter with respect to a reference direction.

struts, which have different thicknesses at the junction between two cells than between three or more cells. The quantification of the latter thus necessitates that an average of the overall cell wall thickness is obtained for each cell considered, which makes image analysis time-consuming and considerably more complex. The second problem originates from the foam surface preparation. Using a microtome to produce foam sections will produce significant deformation of cell walls, either by shear, crushing, or tearing, which will modify the cell wall appearance. Using cryogenic fracture to produce foam sections might result in fracture at the thinnest walls due to their lower resistance. Depending on the properties of the polymer, its mode of fracture in the wall, i.e., whether there is a brittle fracture or ductile tearing with or without local reduction in section (necking), might also modify the cell wall appearance, especially when foams of different polymers are compared.

Because of these difficulties, measurements of the cell wall thickness to quantify the amount of material in Equation 6.6 should be avoided and replaced by a parameter also representing the amount of material: the density. A single microstructural parameter proposed to represent quantitatively the foam morphology [42] is expressed as follows:

$$\rho^* \cdot \frac{1}{\bar{d}} \quad \text{or} \quad \frac{\rho^*}{\bar{d}} \qquad (6.12)$$

In the latter equation, the only assumption made is that the cell size distribution is statistically normal, which in most cases represents the reality since $\bar{d}_v/\bar{d}_n$ is, for all foams considered here, close to 1.

The rationale for using this parameter can be obtained from the cell density in the foams, which is derived from the foam density ρ^*, given by:

$$\rho^* = \frac{m_p}{V_t} = \frac{m_p}{V_p + V_v} = \frac{\rho_p \cdot V_p}{V_p + V_v} \qquad (6.13)$$

where m_p and ρ_p are the polymer mass and density in the foam, and V_t, V_p, and V_v are the total volume of the foam, the volume of the unfoamed polymer, and the volume of the voids. By assuming a unit volume for the unfoamed polymer and spherical cells, the foam density becomes the following:

$$\rho^* = \frac{\rho_p}{1 + V_v} = \frac{\rho_p}{1 + N \pi d^3/6} \qquad (6.14)$$

where N is the number of cells per unit volume, or cell density, and d is their diameter.

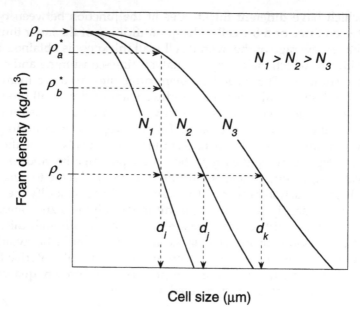

Cell size (μm)

FIGURE 6.6
Graphical representation of foam density as a function of cell density according to Equation 6.15. Three different cell densities, N_1, N_2, and N_3, two specific foam densities, ρ_a^*, ρ_b^*, and ρ_c^*, and three different diameters, d_i, d_j, and d_k, are indicated.

From Equation 6.14, the cell density N can be obtained:

$$N = 6\frac{\left[(\rho_p/\rho^*)-1\right]}{\pi d^3} \approx \frac{6\rho_p}{\pi d^3 \rho^*} \qquad (6.15)$$

Foams can thus be represented morphologically using Equation 6.15, as shown in Figure 6.6, by their density ρ and their cell size d, which define their cell density N.

This figure illustrates the limitations of using the foam density as a unique microstructural parameter. As schematically shown in Figure 6.6, foams of density ρ_c^* with cell densities N_1, N_2, and N_3, such that $N_1 > N_2 > N_3$, can exhibit cell sizes d_i, d_j, and d_k, such that $d_i < d_j < d_k$, respectively. Inversely, foams with densities ρ_a^*, ρ_b^*, and ρ_c^*, such that $\rho_a^* > \rho_b^* > \rho_c^*$, can have similar cell size d_i with different cell densities N_1, N_2, and N_3. This example shows quite clearly that the density alone cannot define foams morphologically such as those presented in Figure 6.6. Their morphology can however be distinctly represented by two independent microstructural parameters: the foam density and the cell size. It is thus proposed that these two parameters be incorporated for convenience of use into a single ratio value, the microstructural parameter $\rho^*/\bar{d}$.

6.5 Morphology-Mechanical Behavior of Foams

6.5.1 Mechanical Behavior vs. Cell Size in a Single Parameter for Polystyrene Foams

The materials used to explore the potential use of the microstructural parameter $\rho^*/\bar{d}$ to relate and report mechanical properties are commercially available PS foam boards from Dow Chemical and Owens Corning. The foams tested had a density ranging between 24.8 and 73.9 kg/m³ with a fully closed-cell structure. Their number-average cell size varied between 78 μm for the highest density and 231 μm for the lowest density. Their cell size dispersity was 1.21 ± 0.03, which indicates that they had a statistically normal cell size distribution. The number-average cell size did not show a systematic correlation with the foam density, as shown in Figure 6.7. The foams studied thus do not represent a family of foams with the same specific morphology or $\rho^*/\bar{d}$, contrary to the foams reported in Section 6.3.2 (Figure 6.4). These foams were subjected to low-speed compression tests, dynamic shock cushioning tests (high-speed compression tests), and falling dart impact tests.

6.5.1.1 Compression Testing

The mechanical properties of the PS foams were obtained [42] from compression testing at low strain rates ($\approx 2 \times 10^{-3}$ s⁻¹). The compression properties

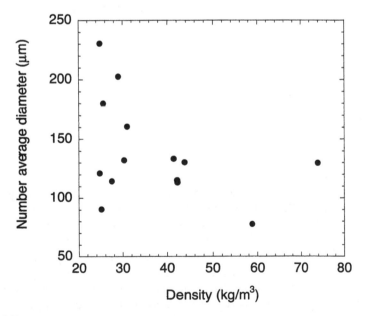

FIGURE 6.7
Number-average cell size plotted against polystyrene foam density. (From Bureau, M.N. and Gendron, R., *J. Cell Plast.*, 39, 353, 2003 [42]. With permission.)

obtained, namely the compressive modulus of elasticity and compressive strength, are plotted against density on a log-log scale in Figure 6.8. This figure shows that both compressive properties can be scaled with density using power-law least-square regressions, as prescribed by Gibson and Ashby, but the power-law exponents obtained from these regressions are not in agreement with Equations 6.3 and 6.4. Given the low standard deviations obtained for these measurements (typically 5% and below) and the fairly good correlation factors obtained from the mechanical data, predicting the compressive properties of the PS foams using Equations 6.3 and 6.4 could lead to an overestimation of the foams in the low density range, as schematically shown by arrows in Figure 6.8. More important, however, is that the Gibson and Ashby approach does not take the foam morphology into consideration, since it only uses the foam density in Equations 6.3 and 6.4. This implies that the density alone governs the mechanical properties of foams, and that the influence of the foam morphology is not significant, which does not represent the properties of foams in reality (see below) [15,29–31,42]. It is thus proposed that a representation of these mechanical properties, taking into account the foam morphology, constitutes in itself an improvement with respect to the classical approach of Gibson and Ashby.

Based on the previous discussion, the mechanical properties are now plotted in Figure 6.9 using the microstructure-density parameter $\rho^*/\bar{d}$ (Section 6.4.5) instead of the density alone on a log-log scale. This figure shows that both compressive properties can be expressed, with reasonably small data deviation, using power-law least-square regressions of the form of Equations 6.3 and 6.4, but with $\rho^*/\bar{d}$ as a foam defining parameter:

$$E^* = c_1 \cdot \left(\frac{\rho^*}{\bar{d}} \right)^{1.4} \tag{6.16}$$

$$\sigma_y^* = c_2 \cdot \left(\frac{\rho^*}{\bar{d}} \right)^{1.0} \tag{6.17}$$

where c_1 and c_2 are proportionality constants.

Given the sources of variation that cell size measurement is subjected to, and given also that the PS foams tested do not present intercorrelated density and cell size (Figure 6.7), the regressions in Figure 6.9 suggest that the mechanical properties of PS foams can be correlated using $\rho^*/\bar{d}$. It is proposed that this approach should be used to correlate properties of foams, since it captures the essential characteristics of the foam morphology and provides property prediction within an acceptable range of error. If demonstrated to be correct, this new representation of mechanical properties would provide more information for the prediction or the interpretation of foam properties than simply the range in density required. This suggests that it

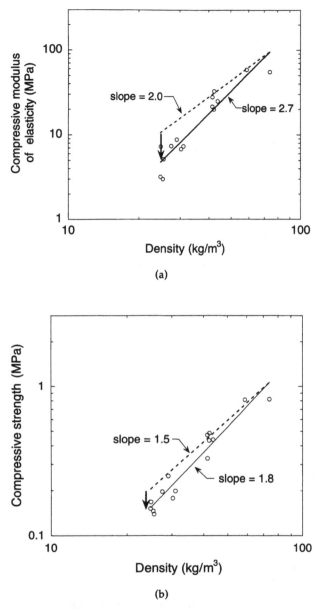

(a)

(b)

FIGURE 6.8
Compressive properties of polystyrene foams plotted against density on a log-log scale: (a) modulus of elasticity ($r^2 = 0.83$) and (b) compressive strength ($r^2 = 0.91$). Dotted lines represent power-law equations prescribed by Gibson and Ashby. Solid lines represent least-square regressions obtained from data. Slope values indicate power-law exponents. Arrows represent the gap at low density between predicted power-law equations and least-square regressions. (From Bureau, M.N. and Gendron, R., *J. Cell. Plast.*, 39, 353, 2003. With permission.)

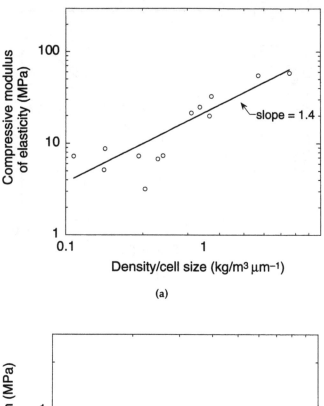

(a)

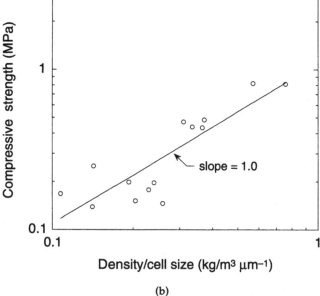

(b)

FIGURE 6.9

Compressive properties of polystyrene foams plotted against density/cell size ratio $\rho^{*}/\bar{d}$ on a log-log scale: (a) modulus of elasticity ($r^2 = 0.77$) and (b) compressive strength ($r^2 = 0.73$). Solid lines represent least-square regressions obtained from data. Slope values indicate power-law exponents. (From Bureau, M.N. and Gendron, R., *J. Cell. Plast.*, 39, 353, 2003 [42]. With permission.)

could be possible to maintain the mechanical properties of a foam while reducing its density by decreasing its average cell size, so that $\rho^*/\bar{d}$ values would remain constant. Such a result would represent obvious economic advantages and also provide support for the idea of MCF (Section 6.3.1). However, this correlation needs to be confirmed over a wider range of $\rho^*/\bar{d}$ values, as only one decade was considered for these PS foams. It also needs to be validated for other testing conditions as well, e.g., falling dart impact or dynamic shock cushioning tests. In addition, it should be extended to other foam materials — polyolefins, for example.

6.5.1.2 *Dynamic Shock Cushioning Testing*

To verify the validity of this method at considerably higher strain rates, the same PS foams were tested by means of a flat sheet impact at high speed (≥ 3 m/s or 120 s^{-1}), as described in Table 6.1. As the dynamic shock cushioning test is meant to measure the amount of energy a material can absorb, structural foams generally necessitate that the test is repeated until the densification stage is reached. When densification occurs, the impact curve of two consecutive tests should be similar. An example of this is given in Figure 6.10. As shown in this figure, the first impact is characterized by a rapid increase of the load until a maximum is reached. Once it is reached, the foam undergoes plastic crushing until the energy of the impactor is totally absorbed. The second impact shows a yield at a considerably lower load, as a result of the significant damage produced during the first impact. This yield is followed by a load increase up to the load level of plastic crushing of the first impact, which extends further until the impact energy is absorbed. The third impact shows only a very slight reduction in yield load, indicating that much of the damage was done during the second impact. Following this yield, the load further increases to levels above those of plastic crushing of the first and second impact, indicating the onset of densification. The curves of the fourth and fifth impacts show the same yield load and the same following increase in load due to densification, with the exception that different maximum densification loads are reached due to the cumulative deformation. Thus, different foams of the same polymeric material, when they are completely crushed and that densification is reached, should present similar impact curves, when the load is normalized for density, as shown in Figure 6.11 for seven different PS foams.

From Figures 6.10 and 6.11, it is not obvious that, for a given level of impact energy, the first impact curves of a series of dynamic shock cushioning tests produce irreversible plastic yielding in the foam. Before it is used to calculate the compressive impact strength, it must be verified that the maximum load from the first impact curve corresponds to plastic yielding in the specimen. A simple way is to verify that the load at yield during the second impact is significantly lower than during the first impact, as a consequence of the damage produced. The compressive impact modulus of elasticity can be obtained from the slope of this impact curve before the first maximum load

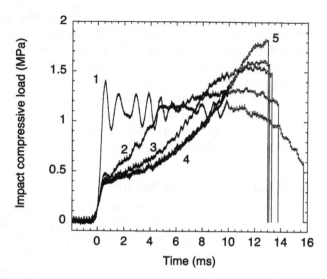

FIGURE 6.10
Example of impact compressive load-time curves of five consecutive dynamic shock cushioning tests on a polystyrene foam. The test numbers are indicated.

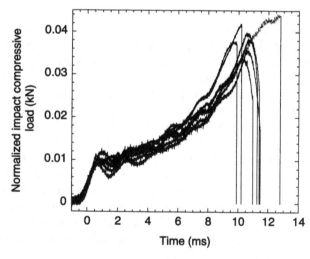

FIGURE 6.11
Normalized impact compressive load-time curves for seven different polystyrene foams (ρ^* between 25.1 and 42.2 kg/m³ and $\rho^*/\bar{d}$ between 0.1 and 0.4 kg/m³ µm⁻¹).

is reached. The compressive impact modulus of elasticity and the compressive impact strength were obtained from the load-deflection curves recorded during the impact tests, as for low-speed compression testing. They are both shown in Figure 6.12 as a function of $\rho^*/\bar{d}$ on a log-log scale.

This figure shows that, as observed from low-speed compressive properties, the compressive impact modulus of elasticity and strength can be scaled

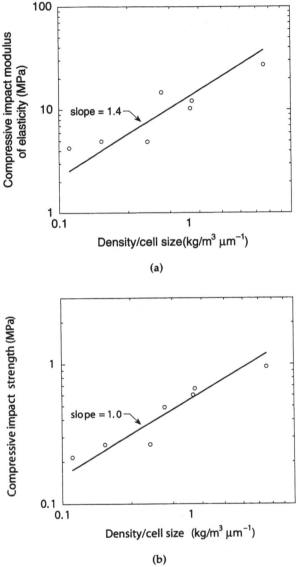

FIGURE 6.12
Compressive impact properties of PS foams plotted against density/cell size ratio $\dot{\rho}^*/\bar{d}$ on a log-log scale: (a) modulus of elasticity ($r^2 = 0.67$) and (b) compressive strength ($r^2 = 0.87$). Solid lines represent least-square regressions obtained from data. Slope values indicate power-law exponents.

with $\rho^*/\bar{d}$, using the power-law regression obtained for low-speed compressive modulus of elasticity (Figure 6.9) and reported in Equations 6.16 and 6.17. Given the four to five orders of magnitude difference in strain rates (2×10^{-3} s^{-1} vs. 120 s^{-1}) in the two tests compared, it is clear that the proposed method can be extended to very wide ranges in testing speed.

6.5.1.3   Falling Dart Impact Tests

To extend the use of the proposed method to other loading conditions, the same PS foams were tested by means of a falling dart at high speed (> 3 m/s), as described in Table 6.1. The impact strength and the impact modulus of elasticity were obtained from the load-deflection curves recorded during the falling dart impact tests. The impact strength and the impact modulus of elasticity were calculated from the Timoshenko–Woinowsky–Krieger equation given for a clamped circular plate loaded at its center [43] using the maximum impact load and the slope, or rigidity, of the load-displacement curve in the elastic regime:

$$\sigma^* = \frac{P}{h^2}\left(1+\upsilon\right)\left(0.485\log\frac{r}{h}+0.52\right) \tag{6.18}$$

$$E^* = \frac{3kr^2}{4\pi h^3}\left(3+\upsilon\right)\left(1-\upsilon\right) \tag{6.19}$$

where P is the load, k is the rigidity, h is the plate thickness, r is the radius of the plate, and υ is the Poisson's ratio of the foam. The falling dart impact strength and modulus of elasticity are shown in Figure 6.13 as a function of $\rho^*/\bar{d}$ on a log-log scale.

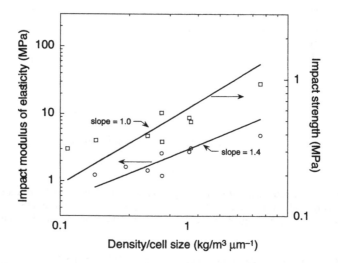

FIGURE 6.13
Falling dart impact modulus of elasticity ($r^2 = 0.62$) and strength ($r^2 = 0.22$) of polystyrene foams plotted against density/cell size ratio $\rho^*/\bar{d}$ on a log-log scale. Solid lines represent least-square regressions obtained from data. Slope values indicate power-law exponents.

This figure shows that, as observed from low- and high-speed compressive properties, the falling dart impact strength and modulus can also be scaled with $\rho^*/\overline{d}$, using the power-law regression obtained for low-speed compressive modulus of elasticity (Figure 6.9) and reported in Equations 6.16 and 6.17. The correlation factor obtained for the falling dart impact strength using the power-law regression of Equation 6.17 was quite poor, however. The complex load modes obtained in such tests at high deformation, such as contact stresses, flexural stresses, compressive stresses, and shear stresses, might be at the origin of the latter. However, given the different strain rates, loading modes, and deformation and fracture mechanisms in low- and high-speed (impact) compression tests and falling dart impact tests, it appears that the proposed microstructural parameter $\rho^*/\overline{d}$, used to express mechanical properties of PS foams with very different microstructural characteristics, can be extended to different testing conditions using the same power-law equations.

6.5.2 Validation of the Single Parameter Approach for Polyolefin Foams

The materials used to validate the latter methodology are commercial polyolefin foams from Dow Chemical (Ethafoams™). These polyolefin foams are essentially composed of different types of polyethylene (PE). These foams had a density between 28.1 and 101.9 kg/m³. One of them had a fully closed-cell structure, while the others had 20–25% open porosity. The cell size distribution was characterized in two planes of observation, one normal to the longitudinal or machine direction and one normal to the thickness of the boards, noted as L and N planes, respectively. The average cell sizes measured in the L planes were generally higher than those in the N planes. The number-average cell size varied between 600 μm for the highest density and 1.79 mm for the lowest density. As observed for the previous PS foams, the cell size dispersity of the polyolefin foams was low, 1.16 ± 0.06, which indicates that they had a statistically normal cell size distribution. These foams were subjected to low-speed compression tests and dynamic shock cushioning tests (high-speed compression tests).

6.5.2.1 Compression Testing

The compressive properties of the PE foams were obtained from compression tests at low strain rates ($\approx 2 \times 10^{-3}$ s^{-1}). The compressive modulus of elasticity and compressive strength are shown in Figure 6.14 as a function of density. Figure 6.14 indicates that both compressive properties can be scaled with density using power-law least-square regressions, though with significant data scatter. In addition, the power-law exponents obtained from the regressions are not in agreement with those prescribed by Gibson and Ashby (Equations 6.3 and 6.4). Given the important data scatter obtained for the modulus and strength measurements, and given that two values of each are reported for a single value of density, i.e., for N and L planes, it appears that the compressive properties of the PE foams tested can not be accurately predicted from Gibson and Ashby, as was concluded for PS foams.

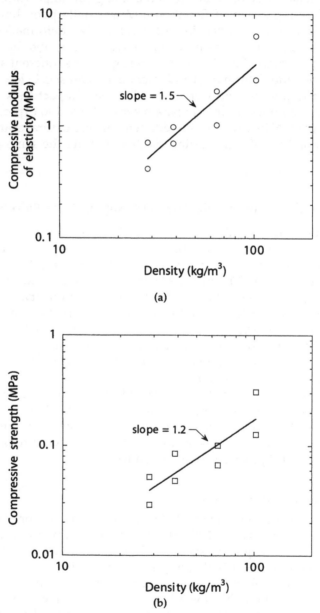

FIGURE 6.14

Compressive properties of polyethylene foams plotted against density on a log-log scale: (a) modulus of elasticity ($r^2 = 0.69$) and (b) compressive strength ($r^2 = 0.63$). Solid lines represent least-square regressions obtained from data. Slope values indicate power-law exponents.

To account for the difference in cell size distribution between foams and for the anisotropy of cells within a foam, the compressive modulus of elasticity and compressive strength are plotted against $\rho^*/\bar{d}$ on a log-log scale in Figure 6.15. This figure shows that both compressive properties can be scaled with $\rho^*/\bar{d}$ using the power-law equations obtained for low-speed

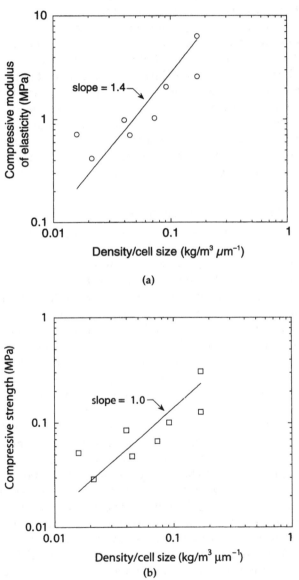

(a)

(b)

FIGURE 6.15
Compressive properties of polyethylene foams plotted against density/cell size ratio $\rho^*/\bar{d}$ on a log-log scale: (a) modulus of elasticity ($r^2 = 0.50$) and (b) compressive strength ($r^2 = 0.54$). Solid lines represent least-square regressions obtained from data. Slope values indicate power-law exponents.

compressive modulus of elasticity of PS foams (Equations 6.16 and 6.17). It thus appears that the relationship between mechanical properties and foam morphology observed for PS foams can be extended to other types of foams, or at least PE foams. However, the deviation from the regression of Figure 6.15 cannot be neglected (e.g., $r^2 \approx 0.50$), which indicates that refinements of the microstructural parameter $\rho^*/\bar{d}$ would be needed.

Unfortunately, the curves obtained using the $\rho^*/\bar{d}$ microstructural parameter did not show a clear improvement in correlation factors. A special feature of the curves in Figure 6.15 could be related to this lack of significant improvements in correlation factor: both curves show what appears as two sets of parallel data. These parallel data sets could be related to an orientation effect in the foams, which was not taken into account when the average cell size was measured. Nevertheless, given the wide range of cell size and density considered for the PE foams, and considering the fact that the mechanical-microstructural relationship observed in PS foams is applied to very different PE foams here, it is reasonable to conclude that the proposed $\rho^*/\bar{d}$ can be used to express the mechanical properties of different foams.

6.5.2.2 Dynamic Shock Cushioning Testing

Dynamic shock cushioning tests were also performed to extend the validity of the method to high strain rates in PE foams. The impact tests were done as described previously for PS foams (Section 6.5.1). The PE foam specimens were subjected to several impacts to ensure that complete densification occurred. The compressive impact curves were recorded and treated as previously explained. However, the compressive impact modulus could not be obtained with acceptable reproducibility due to the low sensitivity of the load cell used at very low loads (e.g., < 100 N), i.e., in the first elastic portion of the curves. The compressive strength, however, could be obtained with good reproducibility. The compressive strength is shown in Figure 6.16 as a function of $\rho^*/\bar{d}$ on a log-log scale. This figure shows that the compressive impact strength can be scaled with $\rho^*/\bar{d}$ using the power-law regression (Equation 6.17). However, important deviation from the power-law regression obtained for PS was observed here for the low $\rho^*/\bar{d}$ values.

The curves of the compressive strengths from low-speed and impact testing (Figures 6.14b and 6.16) show very similar data dispersion. Close and consecutive $\rho^*/\bar{d}$ values gave two significantly different strengths, which explains the dispersion in the data. It is possible that this effect is related to an orientation effect in the foams not taken into account when the average cell size is measured. This orientation or anisotropy effect will be explored in Section 6.5.3. These results nevertheless show that PE foam properties presenting wide ranges in both cell size and density can be represented by the proposed microstructural parameter $\rho^*/\bar{d}$, using the regression parameters obtained from PS foams.

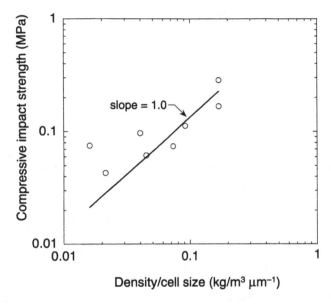

FIGURE 6.16
Compressive impact strength of polyethylene foams from dynamic shock cushioning tests plotted against density/cell size ratio $\rho'/\overline{d}$ on a log-log scale. Solid line represents least-square regression obtained from data ($r^2 = 0.22$). Slope value indicates power-law exponents.

6.5.3 Validation of Single Parameter Approach for Anisotropic Polyolefin Foams

To account for specimen variation and foam anisotropy, PE foams were obtained from Sealed Air Corporation with the same density but very different cell size distributions. These PE foams had a density of 35.3 ± 0.3 kg/m³ and a fully closed-cell structure. The cell size distribution was characterized in three orthogonal planes of observation in the boards, one normal to the longitudinal or machine direction, noted as the L plane; one normal to the transverse direction, noted as the T plane; and one normal to the thickness direction, noted as the N plane. The average cell sizes measured in the N plane were in all cases significantly smaller than those measured in the T and L planes. Close values of average cell size were measured in the T and L planes, but the average cell sizes in the L plane were generally higher than those in the T plane. The cells also appeared more elongated in the L direction than in the T and N directions. The number-average cell size varied between 0.89 mm for the highest density and 1.88 mm for the lowest density. A fairly low cell size dispersity of 1.32 ± 0.05 was measured for these polyolefin foams. Although slightly higher than those measured for the previous PS and PE foams, their cell size dispersity still indicates that they were close to a statistically normal cell size distribution. These foams were subjected to low-speed compression tests and dynamic shock cushioning tests (high-speed compression tests).

6.5.3.1 Compression Testing

The compressive properties of the PE foams were obtained from compression tests under the same conditions as those used for the previous PS and PE foams. The PE foams were tested in the three orthogonal directions. A typical example of the compressive stress–strain curves is shown in Figure 6.17. In agreement with the cell size measurements, these compressive curves reveal the anisotropic character of the PE foams, which translates into different properties in different directions of testing. The compressive stress–strain curves of the foams in the N plane were above those in the L and T planes. While the curves in the L and T planes were closer, the curves in the L plane were generally below those of the T plane. Depending on the orthogonal direction considered, the curves either showed a definite yield point, characterized by a local maximum in stress, or a progressive yield point, characterized by a change of slope from the elastic stage toward the plastic crushing stage (see Section 6.2.1). The progressive yield point was determined on the curve at the compressive stress corresponding to the compressive strain at the intersection between the extrapolation of the slopes in the elastic and plastic crushing stages, as schematized in Figure 6.17. This method was believed to better capture the yield point in the curve than taking the compressive stress at 10%, as recommended by American Society for Testing and Materials (ASTM) D-1621 (see Section 6.2.1), since the progressive yield point occurred well below a compressive strain of 10% in all cases considered.

From these compressive stress–strain curves, the compressive modulus of elasticity and compressive strength were obtained. These compressive properties are shown in Figure 6.18 as a function of density, following the Gibson and Ashby approach. The results in Figure 6.18 underline quite clearly the previous conclusion that the properties of foams cannot be accurately correlated using density, since density does not account for the specific microstructural characteristics of the foams considered. In Figure 6.18, the values of compressive modulus of elasticity and compressive strength vary from 0.5 MPa to 2.0 MPa and from 0.025 MPa to 0.06 MPa, respectively, i.e., variations of +140% to +300% depending on the value considered, for a density that does not vary by more than 9%. These small density variations obviously cannot explain such important variations in compressive properties.

To account for these variations in compressive properties, the compressive modulus of elasticity and compressive strength are plotted against $\rho^*/\bar{d}$ on a log-log scale in Figure 6.19. This figure shows that both compressive properties can be scaled with $\rho^*/\bar{d}$ using the power-law equations (Equations 6.16 and 6.17). However, it also shows important deviation in the data from the regressions of the compressive modulus and strength. Careful examination

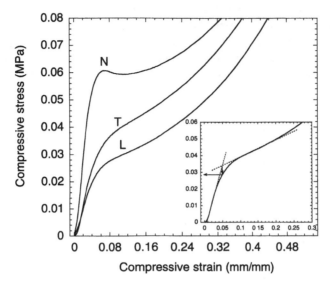

FIGURE 6.17
Typical compressive stress–strain curves for a polyethylene foam in three orthogonal directions (ρ^* of 35.0 kg/m³; $\bar{d}_n$ of 1.05; 1.49 and 1.60 mm for the N, T, and L directions, respectively). An enlarged portion of the L curve is shown in order to represent the progressive yield point that is being considered.

of the curves in Figure 6.19 reveals that most of the scatter in the data is related to the L plane results, while the N and T plane results show less scatter with respect to the power-law regressions. Since the calculation of $\rho^*/\bar{d}$ only considers the equivalent cell size, the effect of cells with different shape factors, i.e., spherical vs. ellipsoidal, and different orientations with respect to the N, L, and T directions, is not reflected in the $\rho^*/\bar{d}$ value. According to the $\rho^*/\bar{d}$ parameter, foams with very elongated cells in the testing direction are predicted to show the same properties as foams with spherical cells, as long as their density and equivalent cell size are the same. It is obvious that the latter statement does not reflect the reality and that an improved $\rho^*/\bar{d}$ parameter should capture the average shape and orientation of cells as well.

6.5.3.2 Correction of the Microstructural Parameter for Anisotropy

By means of image analysis, two parameters of the cell size distribution can easily be obtained in addition to the average cell size and cell size dispersity. The first parameter is the average cell shape F_u in the plane of observation u, given by

$$F_u = \frac{\bar{d}_{u,\text{max}}}{\bar{d}_{u,\text{min}}} \qquad (6.20)$$

where $\bar{d}_{u,max}$ and $\bar{d}_{ui,min}$ are the average maximum cell diameter and the average minimum cell diameter, respectively.

The second parameter is the average cell orientation θ_u, defined as the average angle between the axis of maximum cell diameter and the testing direction in the plane of observation u. These two microstructural parameters

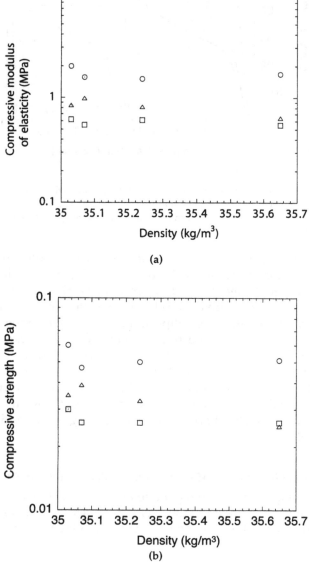

(a)

(b)

FIGURE 6.18

Compressive properties of polyethylene foams plotted against density on a semi-log scale: (a) modulus of elasticity and (b) compressive strength. Results in the *N*, *L*, and *T* planes are represented by ○, □, and △, respectively.

are converted into an anisotropy factor w_u, and applied to the average cell size $\bar{d}_u$. The anisotropy factor w_u is given by

$$w_u = F_u^* \cdot \theta_u^* \tag{6.21}$$

where the subscript u designates the plane considered (N, L, or T), F_u^* and θ_u^* are the calculated cell shape and cell orientation parameters for one plane, given by

$$F_u^* = \frac{F_v \cdot F_w}{\bar{F}} \tag{6.22}$$

$$\theta_u^* = 1 + \frac{\left(\pi/2 - \theta_v\right)}{\pi/2} \cdot \frac{\left(\pi/2 - \theta_w\right)}{\pi/2} \tag{6.23}$$

where the subscripts u, v, and w designate N, L, and T planes, and $\bar{F}$ is the average value of F_N, F_L, and F_T.

Similar to the use of void aspect ratio and void fraction in theoretical calculations of foam elastic properties [29,30] based on Mori-Tanaka, it is possible to express the properties obtained for a given plane u by incorporating the anisotropy factor w_u into the value of $\bar{d}$. The corrected microstructural parameter $\left(\rho^*/\bar{d}\right)_u$ thus becomes

$$\left(\rho^*/\bar{d}\right)_u = \frac{\rho^*}{\bar{d} \cdot w_u} \tag{6.24}$$

The compressive modulus of elasticity and compressive strength previously presented in Figure 6.19 are now plotted against $\left(\rho^*/\bar{d}\right)_u$ on a log-log scale in Figure 6.20. As shown in Figure 6.20, using the corrected $\left(\rho^*/\bar{d}\right)_u$ parameter for foam anisotropy results in significant improvement in data scatter in comparison to Figure 6.19, which simply used $\rho^*/\bar{d}$. Using the power-law regressions (Equations 6.16 and 6.17) also resulted in considerably improved correlation factors ($r^2 \approx 0.6$–0.7) with respect to those obtained in Figure 6.19 ($r^2 \approx 0.2$–0.3). While the deviations of the data from the regressions in Figure 6.20 are still present, their improvement with respect to the deviations in Figure 6.19 suggests that the correction proposed captures the anisotropic nature of the foams studied. The assumptions made in the computation of elastic properties of foams using Mori-Tanaka [29,30], according to which the voids in a foam can be considered as second phase particles with elastic properties and a Poisson's ratio of 0, thus appears to be valid, since elastic properties determined along and normal to elongated foam cells vary accordingly. It also indicates that, as for all foams in all testing conditions reported here, the foams follow the same power-law equations (Equations 6.16 and 6.17) with respect to modulus of elasticity and strength.

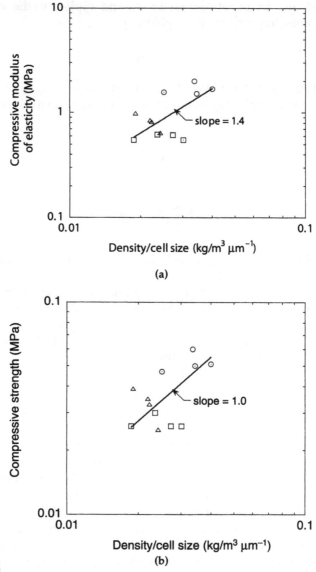

FIGURE 6.19

Compressive properties of polyethylene foams plotted against density/cell size ratio $\rho / \bar{d}$ on a log-log scale: (a) modulus of elasticity ($r^2 = 0.32$) and (b) compressive strength ($r^2 = 0.24$). Results in the N, L, and T planes are represented by $\bigcirc$, $\square$, and $\triangle$, respectively. Solid lines represent least-square regressions obtained from data. Slope values indicate power-law exponents.

6.5.3.3   *Dynamic Shock Cushioning Testing*

These PE foams were also subjected to dynamic shock cushioning tests. The impact tests were done as described in Section 6.5.2. As previously mentioned, only the compressive strength could be obtained with acceptable

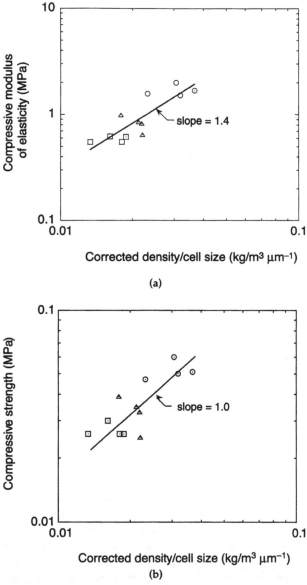

FIGURE 6.20
Compressive properties of polyethylene foams plotted against the corrected density/cell size ratio $(\rho^*/\bar{d})$, on a log-log scale: (a) modulus of elasticity ($r^2 = 0.72$) and (b) compressive strength ($r^2 = 0.60$). Results in the N, L, and T planes are represented by $\bigcirc$, $\square$, and $\triangle$, respectively. Solid lines represent least-square regressions obtained from data. Slope values indicate power-law exponents.

levels of reproducibility. The compressive impact strength is shown in Figure 6.21 as a function of $\rho^*/\bar{d}$ on a log-log scale. As observed for the low-speed compression tests, Figure 6.21 shows that compressive impact strength can be scaled with $\rho^*/\bar{d}$ using equations of the form of

Equation 6.17, but contrary to the low-speed compression test results, the exponent obtained from the regression is 0.7, lower than the exponents of 1.0 obtained for low-speed compressive modulus of elasticity of PS foams (Figure 6.9) and reported in Equations 6.16 and 6.17. Similar to Figure 6.19, an important deviation of the data from the regression of the compressive strength was noted in Figure 6.21. In agreement with observations in Figure 6.19, the data scatter appears to be related mostly to the L plane results, while the N and T plane results show less scatter with respect to the power-law regression employed. The correct $\left(\rho^*/\bar{d}\right)_u$ parameter for foam anisotropy is thus applied to the results of Figure 6.21 to account for the aspect ratio and orientation of the cells not reflected in the microstructural parameter $\rho^*/\bar{d}$. The compressive strength is plotted against $\left(\rho^*/\bar{d}\right)_u$ on a log-log scale in Figure 6.22. As observed for the low-speed compression tests, using the corrected $\left(\rho^*/\bar{d}\right)_u$ parameter in Figure 6.22 for foam anisotropy results in significant improvement in data scatter in comparison to data scatter in Figure 6.21, which was based on the uncorrected $\rho^*/\bar{d}$ parameter. Also as observed for the low-speed compression tests, Figure 6.22 shows that compressive impact strength can be scaled with a modified $\rho^*/\bar{d}$ using equations of the form of Equation 6.17, with the same power-law exponent of 0.7 as obtained in Figure 6.21. The results from compression testing of PE foams at low and high speeds, presenting a similar density but a wide range in cell size and thus very different values of $\rho^*/\bar{d}$, indicate that the incorporation of a cell shape factor (aspect ratio) and a cell orientation factor into the microstructural parameter $\left(\rho^*/\bar{d}\right)_u$ permits us to capture the anisotropic nature of foams. They also indicate that the trends observed in power-law regressions can be applied to different materials at different strain rates.

6.5.4 Discussion

To obtain a global view of the mechanical properties of the PS and PE foams considered here, all data on modulus of elasticity and strength obtained are plotted against $\rho^*/\bar{d}$ on a log-log scale in Figure 6.23. A first observation from this figure is that the range considered in mechanical properties as well as in microstructural parameter $\rho^*/\bar{d}$ extends over two decades for the strength and three for the modulus, which shows how diversified the properties and materials considered are: semicrystalline and amorphous polymeric foams, with densities between 20 and 100 kg/m³, cell sizes between 0.078 mm and 1.9 mm, and various levels of anisotropy. Using the regressions from Equations 6.16 and 6.17 for the modulus and strength, and following the power-law exponents of 1.4 and 1.0, respectively, for the modulus and strength obtained initially for PS foams (Section 6.5.1), it appears that all results lie along a single straight line of data, with very good regression factors ($r^2 = 0.80$), considering the diversity in properties and materials

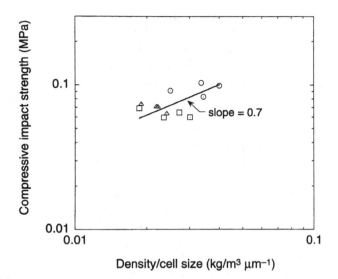

FIGURE 6.21
Compressive impact strength of polyethylene foams from dynamic shock cushioning tests plotted against density/cell size ratio $\rho^{\cdot}/\bar{d}$ on a log-log scale. Results in the N, L, and T planes are represented by ○, □, and △, respectively. The solid line represents least-square regressions obtained from data ($r^2 = 0.50$). The slope value indicates the power-law exponent.

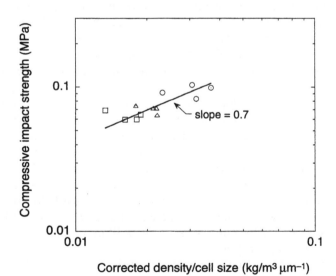

FIGURE 6.22
Compressive impact strength of polyethylene foams from dynamic shock cushioning tests plotted against corrected density/cell size ratio $(\rho^{\cdot}/\bar{d})_u$ on a log-log scale. Results in the N, L, and T planes are represented by ○, □, and △, respectively. The solid line represents least-square regressions obtained from data ($r^2 = 0.53$). The slope value indicates the power-law exponent.

considered. This observation clearly demonstrates that the mechanical properties of polymeric foams can be expressed as a function of their microstructural characteristics, using the $\rho^*/\bar{d}$ parameter. The results from the corrected $\rho^*/\bar{d}$ parameter for anisotropy indicate, however, that cell aspect ratio and orientation should be considered for expressing the average cell size, following Equation 6.24. The latter were only considered for the PE foams treated in Section 6.5.3.

In light of the success of this approach, some general guidelines for foam processing might be suggested. The first is that the power-law regressions proposed in Equations 6.16 and 6.17 could be used to optimize the properties of foams. The latter equations indicate that the properties of a foam at a given density could be enhanced by a reduction of the average cell size, or that the properties could be maintained at a lower density by reducing the cell size, to keep the microstructural parameter $\rho^*/\bar{d}$ constant. This indication provides support to MCFs, which show properties claimed to be maintained with respect to the unfoamed polymer for typical density reductions of 30%, as a result of a very fine cellular structure. Extrapolation, schematized in Figure 6.24, of the global regressions for modulus and strength in Figure 6.23 indicates that for typical values for MCFs (see Section 6.3.1), i.e., relative densities of 0.7 to 0.8 and average cell size of 10 to 40 µm, the modulus of elasticity predicted is between 4 and 10 GPa, and the strength between 30 and 60 MPa, which appear to be in the right range of properties (Section 6.3.1). Of course, more work needs to be done on specific polymer foams for different ranges in density (e.g., relative densities between 0.3 to 0.8) before a definitive conclusion can be reached on the validity of this extrapolation to the whole density and microstructure spectrum, but the extrapolated ranges in Figure 6.24 at least are in agreement with observed results for MCFs.

Provided that it is validated for a specific system, the correlation between the properties and the microstructural parameter $\rho^*/\bar{d}$ also indicates an opportunity in terms of costs in materials, since the same properties could be obtained for a lower density, i.e., for a smaller amount of material. This effect of microstructure is not predicted in the Gibson and Ashby relationship (see Section 6.2.2), since this relationship only considers the relative density, and thus predicts that a smaller amount of materials leads to lower properties. However, the latter should be considered as a macroscopic tool for comparison between different foams, or even different materials, and not as a means of predicting properties.

Another guideline is that anisotropy could be used as a tool to adjust specific properties in specific directions while maintaining foam density at a given level. The corrected microstructural parameter $\left(\rho^*/\bar{d}\right)_u$ or $\rho^*/(\bar{d}\cdot w_u)$ suggests that enhanced properties could be obtained at lower $\bar{d}\cdot w_u$ values. For a given density, higher properties in a direction x thus could be obtained by producing a foam presenting a lower cell size with a smaller anisotropy factor w_u in the tested plane normal to direction x, e.g.,

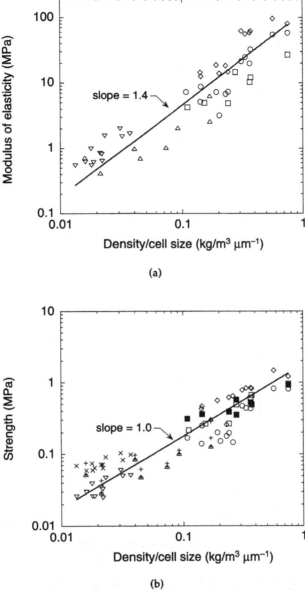

(a)

(b)

FIGURE 6.23
Modulus of elasticity and strength plotted against density/cell size ratio $\rho^{*}/\bar{d}$ for polystyrene foams from compression (O), falling dart impact (■), dynamic shock cushioning (□), and three-point bending (◇), and for polyethylene foams from two sets of compression test data (△: first batch of foams, anisotropy not considered, ▽: second batch of foams, anisotropy considered) and of dynamic shock cushioning test data (+: first batch of foams, anisotropy not considered, ×: second batch of foams, anisotropy considered). The solid line represents least-square regressions obtained from data ($r^2 = 0.78$ for the modulus and $r^2 = 0.83$ for the strength). Slope values indicate power-law exponents.

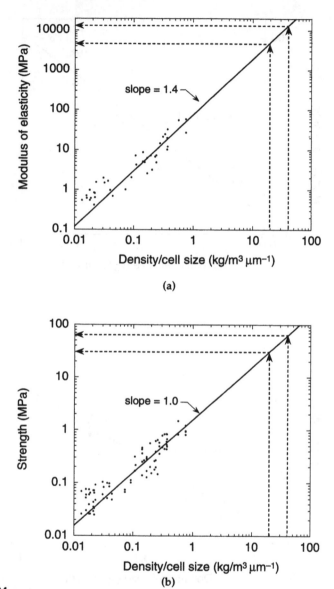

FIGURE 6.24
Extrapolation of regressions for modulus and strength in Figure 6.23 to higher values of $\rho^{*}/\overline{d}$. Dashed lines represent ranges of modulus, strength, and $\rho^{*}/\overline{d}$ values.

by creating an oriented structure of elongated cells in the direction parallel to the plane of testing ($\theta \approx 0$ and $F > 1$ in direction x). From the suggested effect, this oriented cellular structure would behave as predicted by theoretical calculations of a foam's elastic properties [29,30] based on Mori-Tanaka, which considers the internal stress in the matrix of a material containing second phase particles, assimilated to voids for foams, and takes into account their size and aspect ratio.

6.6 Summary and Conclusions

The objective of this chapter was to study the mechanical properties of thermoplastic foams in view of their morphology, defined not only by their density but also by their cell size distribution. Classical mechanical properties used to characterize the performance of foams were explored. Compression testing, both at high (impact) and low strain rates, appears to be the most popular mechanical characterization technique, although other techniques have been employed. The literature review showed that the predominant approach used to rationalize the mechanical properties of foams is based on the work of Gibson and Ashby, who related properties to density using the optimization theory for materials selection in design. This approach considers a simple model consisting of a cubic array of rigid members (struts) and cell walls for closed-cell foams from which the basic mechanical properties can be obtained as a function of the density. Although this approach is quite successful in relating properties of very different classes of materials in terms of their density, the approach is highly macroscopic and does not consider the particularities of each class of materials with respect to their morphology and architecture.

Indeed, the literature review showed that the specific morphological characteristics of MCFs have a great influence on their mechanical properties, and that the cell size, among other microstructural parameters, plays an important role in the determination of these properties. Different studies based on continuum mechanics approaches and finite element modeling indicated that not only the void content (density), but also their orientation and aspect ratio, affected the elastic properties of foams. Experimental studies led, however, to inconsistent results concerning the effect of specific morphological characteristics of foams, namely their cell size and cell wall thickness, on their mechanical properties. In some cases, clear evidence of an improved performance for lower cell size was reported, whereas in others, no effect of cell size or cell wall thickness on mechanical properties was noted. Our own analysis of some works reported in the literature showed, however, that results might have been misinterpreted.

It was thus felt that refinements to the Gibson and Ashby approach, with respect to the morphology of foams, would be appropriate. Specific useful morphological characteristics, in addition to the density, were identified: the type of structure (closed vs. open), the cell size distribution (number- and volume-average diameters and cell size dispersity), and the foam anisotropy (cell size distribution, average cell aspect ratio, and cell orientation in different planes of observation). A microstructural parameter could be derived from the relationship between the cell density, the average cell size, and the density of the foamed and unfoamed polymer. This parameter, given by the ratio between the average cell size and the foam density ($\rho^*/\bar{d}$), is proposed to represent the foam morphology specifically.

The use of this parameter was then validated for three series of different foams, one series of PS foams and two different series of polyolefin foams. Morphological and mechanical characterization of these three series of foams showed that their mechanical properties can be correlated using $\rho^*/\bar{d}$, in most cases with reasonably low data deviation. The case of the first series of PE foams indicated that the anisotropic nature of the foams had to be considered in the property–microstructure ($\rho^*/\bar{d}$) correlation, the properties obtained in the N plane being generally above those in the L plane, which led to an increase of data deviation from the property–microstructure ($\rho^*/\bar{d}$) correlation. The second series of PE foams showed that the mechanical properties of foams with very close densities (35.3 ± 0.3 kg/m³) systematically ranged with the cell size, but acceptable data deviation could not be obtained using the property–microstructure ($\rho^*/\bar{d}$) correlation. The microstructural parameter $\rho^*/\bar{d}$ was modified to include the aspect ratio of cells (F_u) and their orientation (θ_u) with respect to the testing direction, in agreement with elastic property computation using Mori-Tanaka [29,30]. This corrected microstructural parameter $\left(\rho^*/\bar{d}\right)_u$ for foam anisotropy was defined as $\rho^*/\bar{d} \cdot w_u$, where w_u is a function of F_u and θ_u. Using the corrected microstructural parameter $\left(\rho^*/\bar{d}\right)_u$ proved to improve the property–microstructure correlation.

A striking observation from this newly proposed property–microstructure correlation, whether $\rho^*/\bar{d}$ or $\left(\rho^*/\bar{d}\right)_u$ is used, is that not only can the mechanical properties of very different foams with very different morphologies be correlated using the proposed property–microstructure relationship, but also all mechanical data obtained from these three series of foams using this microstructural parameter lie on the same property–microstructure correlation. In light of this successful correlation, it must be concluded that this new representation of mechanical properties is valid and provides more information for the prediction or the interpretation of foam properties than the classical Gibson and Ashby approach, since it defines two of the three essential morphological features of foams — the foam density, the cell size, and the cell density — instead of only the density in the case of the classical approach. It also must be concluded that the combination of these two morphological features, the foam density and average cell size, into a single microstructural parameter, whether $\rho^*/\bar{d}$ or $\left(\rho^*/\bar{d}\right)_u$, captures the essential characteristics of the foam morphology, at least for PS and PE foams developed for commercial applications. This newly proposed property–microstructure correlation indicates that foams properties can be maintained at lower densities by decreasing average cell size, so that $\rho^*/\bar{d}$ values remain constant, or inversely that foam properties could be improved by decreasing average cell size without increasing density.

Abbreviations

ASTM:	American Society for Testing and Materials
DMA:	Dynamic mechanical analysis
EVA:	Polyethylene vinyl acetate
HDPE:	High-density polyethylene
IS:	Isoprene–styrene copolymer
LDPE:	Low-density polyethylene
LLDPE:	Linear low-density polyethylene
MCFs:	Microcellular foams
PC:	Polycarbonate
PE:	Polyethylene
PS:	Polystyrene
SAN:	Styrene–acrylonitrile copolymer

References

1. Gibson, L.J. and Ashby, M.F., *Cellular Solids: Structure and Properties*, 2nd ed., Cambridge University Press, Cambridge, U.K., 1997.
2. American Society for Testing and Materials, ASTM D1621-00, Standard Test Method for Compressive Properties of Rigid Cellular Plastics.
3. American Society for Testing and Materials, ASTM C203-99, Standard Test Methods for Breaking Load and Flexural Properties of Block-Type Thermal Insulation.
4. American Society for Testing and Materials, ASTM D5628-96, Standard Test Method for Impact Resistance of Flat, Rigid Plastic Specimens by Means of a Falling Dart.
5. American Society for Testing and Materials, ASTM D1596-97, Standard Test Method for Dynamic Shock Cushioning Characteristics of Packaging Material.
6. Ashby, M.F., *Materials Selection in Mechanical Design*, Butterworth-Heinmann, Oxford, U.K., 1992, 311 p.
7. Martini, J.E., Suh, N.P., and Waldman, F.A., U.S. Patent 4,473,665, 1984.
8. Kumar, V. and Suh, N.P., A process for making microcellular thermoplastic parts, *Polym. Eng. Sci.*, 30, 1323, 1990.
9. Kumar, V. et al., Experimental characterization of the tensile behavior of microcellular polycarbonate foams, *J. Eng. Mater. Technol.*, 116, 439, 1994.
10. Kumar, V., Weller, J.E., and Murray, R., Microcellular ABS foams, *Proc. SPE Annu. Mtg. ANTEC*, 2202, 1995.
11. Wing, G. et al., Time dependent response of polycarbonate and microcellular polycarbonate, *Polym. Eng. Sci.*, 35, 673, 1995.
12. Weller, J.E. et al., The tensile properties of microcellular PVC: the effects of additives and foam density, *Proc. SPE Annu. Mtg. ANTEC*, 2055, 1997.

13. Barlow, C.C. et al., Experiments on the impact strength of microcellular poly-carbonate, in *Proc. ASME Porous, Cell. Microcell. Mater.*, MD-82, 1998, 45.
14. Barlow, C.C. et al., Some experiments on the fracture behavior of microcellular polycarbonate, in *Proc. ASME Porous, Cell. Microcell. Mater.*, MD-82, 1998, 35.
15. Arora, K.A., Lesser, A.J., and McCarthy, T.J., Compressive behavior of micro-cellular polystyrene foams processed in supercritical carbon dioxide, *Polym. Eng. Sci.*, 38, 2055, 1998.
16. Sun, H., Sur, G.S., and Mark, J.E., Microcellular foams from polyethersulfone and polyphenylsulfone, preparation and mechanical properties, *Eur. Polym. J.*, 38, 2373, 2002.
17. Sun, H. and Mark, J.E., Preparation, characterization, and mechanical proper-ties of some microcellular polysulfone foams, *J. Appl. Polym. Sci.*, 86, 1692, 2002.
18. Lin, H.-R., The effects of foaming and annealing on mechanical properties and DSC scans of gas saturated and non-gas saturated high impact polystyrene, *Cell. Polym.*, 17, 402, 1998.
19. Vanvuchelen, J. et al., Microcellular PVC foam for thin wall profile, *J. Cell. Plast.*, 36, 148, 2000.
20. Collias, D.I., Baird, D.G., and Borggreve, R.J.M., Impact toughening of poly-carbonate by microcellular foaming, *Polymer*, 35, 3978, 1994.
21. Collias, D.I. and Baird, D.G., Tensile toughness of microcellular foams of poly-styrene, styrene-acrylonitrile copolymer, and polycarbonate, and the effect of dissolved gas on the tensile toughness of the same polymer matrices and microcellular foams, *Polym. Eng. Sci.*, 35, 1167, 1995.
22. Collias, D.I. and Baird, D.G., Impact behavior of microcellular foams of poly-styrene, styrene-acrylonitrile copolymer, and single-edge-notched tensile toughness of microcellular foams of polystyrene, styrene-acrylonitrile copoly-mer, and polycarbonate, *Polym. Eng. Sci.*, 35, 1178, 1995.
23. American Society for Testing and Materials, ASTM D5045-99, Standard Test Methods for Plane-Strain Fracture Toughness and Strain Energy Release Rate of Plastic Materials.
24. American Society for Testing and Materials, ASTM E1820-01, Standard Test Method for Measurement of Fracture Toughness.
25. American Society for Testing and Materials, ASTM D6110-02, Standard Test Methods for Determining the Charpy Impact Resistance of Notched Specimens of Plastics.
26. American Society for Testing and Materials, ASTM D256-02, Standard Test Methods for Determining the Izod Pendulum Impact Resistance of Plastics.
27. Kumar, V., Juntunen, R.P., and Barlow, C., Impact strength of high relative density solid state carbon dioxide blown crystallizable poly(ethylene tereph-thalate) microcellular foams, *Cell. Polym.*, 19, 25, 2000.
28. Schjodt-Thomsen, J. and Pyrz, R., Effective properties of cellular materials, *Polym. Eng. Sci.*, 41, 752, 2001.
29. Chao, L.-P. and Huang, J.H., Prediction of elastic moduli of porous materials with equivalent inclusion method, *J. Reinf. Plast. Compos.*, 18, 592, 1999.
30. Chao, L.-P., Huang, J.H., and Huang, Y.-S., The influence of aspect ratio of voids on the effective elastic moduli of foamed metals, *J. Compos. Mater.*, 33, 2002, 1999.
31. Kim, D.W. and Kim, K.S., Investigation of the radiation crosslinked foams produced from metallocene polyolefin elastomers/polyethylene blend, *J. Cell. Plast.*, 37, 333, 2001.

32. Rodriguez-Perez, M.A., Rodriguez-Llorente, S., and De Saja, J.A., Dynamic mechanical properties of polyolefin foams studied by DMA techniques, *Polym. Eng. Sci.*, 37, 959, 1997.
33. Velasco, J.I. et al., Application of instrumented falling dart impact to the mechanical characterization of themoplastic foams, *J. Mater. Sci.*, 34, 431, 1999.
34. Rodriguez-Perez, M.A. and De Saja, J.A., The effect of blending on the physical properties of crosslinked closed cell polyethylene foams, *Cell. Polym.*, 18, 1, 1999.
35. Velasco, J.I. et al., Rigidity characterization of flexible foams by falling dart rebound tests, *Cell. Polym.*, 19, 115, 2000.
36. Rodriguez-Perez, M.A. et al., Mechanical characterization of closed-cell polyolefin foams, *J. Appl. Polym. Sci.*, 75, 156, 2000.
37. Almanza, O., Rodriguez-Perez, M.A., and De Saja, J.A., The microstructure of polyethylene foams produced by a nitrogen solution process, *Polymer*, 42, 7117, 2001.
38. Rodriguez-Perez, M.A., The effect of chemical composition, density and cellular structure on the dynamic mechanical response of polyolefin foams, *Cell. Polym.*, 21, 117, 2002.
39. Rodriguez-Perez, M.A. et al., The effect of cell size on the physical properties of crosslinked closed cell polyethylene foams produced by a high pressure nitrogen solution process, *Cell. Polym.*, 21, 165, 2002.
40. Saltikov, S.A., *Stereometric Metallography*, 2nd ed., Metallurgizdat, Moscow, 1958.
41. Underwood, E.E., Particle- and grain-size distributions, in *Quantitative Stereology*, Addison-Wesley, Reading, MA, 1970, chap. 5.
42. Bureau, M.N. and Gendron, R., Mechanical-morphology relationship of PS foams, *J. Cell. Plast.*, 39, 353, 2003.
43. Timoshenko, S. and Woinowski-Krieger, S., *Theory of Plates and Shells*, McGraw-Hill, 1959, 67.

Index

A

Abrasion resistance, 119
Acrylonitrile, 125–126
Acrylonitrile–butadiene–styrene (ABS)
    blends with, 109, 124, 131
    rheological behavior, 74
    solubility and diffusivity in, 24–25
Acrylic processing aids, 65–67
Acrylic, in foamable polymer blends, 121–123
Activation energy
    diffuse, 8, 10
    for viscosity, *see* Energy of activation of
        flow
    permeability, 15
Activity, penetrant, 11, 12, 14, 29
Additives, 71, 124, 162, 169–170
    carbonyl-containing, 153
    chemical, 2
    CO_2-philic, 125
    Organically modified clay, 9
    oxygen-containing, 152
    polymers as, 106
    titanium dioxide, 9
Aging, foam, 30, 31, 147
Alcohols,
    ethanol, 178
    isopropanol, 158, 160, 174–175, 176,
        177–178
    2-ethyl-hexanol, 81–83, 158, 160, 186
American Society for Testing and Materials,
    171, 266
Ammonia, 142
Amorphous polymers
    foam stabilization with, 97, 156
    gas sorption in, 4
    glass transition temperature of, 34, 118
    in polymer blends, 109, 124
    viscosity behavior of, 48, 49, 58, 69, 75, 96
Amorphous phase
    gas sorption and transport in, 7, 8, 12
    in poly(vinyl chloride), 64–65
Anisotropy, cells
    characterization of, 250, 265
    factor, 269, 278

impact on mechanical properties, 250,
    264, 272, 274–275, 277
Argon (Ar)
    in foam processing, 147
    physical properties of, 150
    solubility and diffusivity of, 7, 9, 18, 25
Arrhenius-type equation, 49
ASTM, *see* American Society for Testing and
    Materials
Atmospheric gases, 117, 187; *see also* Carbon
    dioxide; Nitrogen; Oxygen
Attenuation of ultrasonic waves
    definition and computation of, 200, 201,
        202, 204
    response to chemical blowing agent
        decomposition, 226–227, 229
    response to degassing, 207, 208,
    response to plasticization, 206, 210–211
    vs. dispersion of blowing agent, 168–169,
        177
Autoclave foam process, *see* Batch foaming
Azodicarbonamide (AZ), *see* Chemical
    blowing agent

B

Bagley plot or correction, 50, 56
Batch foaming, 26, 27, 29, 69, 121, 153, 154
Bernstein-Kearsley-Zapas (BKZ) model, 56
Biaxial extension or stretching, 52, 53, 54, 63,
    85
    in cell growth, 52, 60, 75
Biodegradable polymers, 10, 109
Biurea, 163
Block copolymers
    in polymer blends, 126, 127, 128
    rheological behavior of, 74
Blowholes, 160, 225
Blow molding, 52, 53, 85, 87
Blowing agents, *see* Chemical blowing
        agents; Physical foaming agents;
        Physical foaming agent mixtures
Blowing power, 29, 146, 147, 158, 159, 162, 181
Board, foam, 31, 154, 156, 160, 253
Boiling point

Hydrochlorofluorocarbons (HCFC)
  damaging the ozone layer, 2, 143
  for foaming, 117, 132, 145, 147, 154, 160
  HCFC-22
    in foam processing, 122, 124, 145, 160
    physical properties, 144
    solubility, 25, 177
  HCFC-142b
    degassing, 207, 208, 209, 210,
    in foam processing, 31, 122, 145, 156,
      160, 198
    physical properties of, 144
    plasticizing effect, 75–77, 79, 80–82,
      173, 212–213
    solubility of, 13, 25, 150
  nomenclature, 145
  physical properties, 144, 146
Hydrodynamic volume, 69
Hydrofluorocarbons (HFC)
  as replacement for HCFC, 145, 154, 156,
    160
  HFC-134a
    degassing of, 177–179, 213–214,
      220–222
    expansion profiles of, 181, 183
    impact on nucleation density, 156–157,
      183–186, 222–224
    in foam processing, 31, 96, 122–123,
      156, 157, 158, 183, 205
    physical properties of, 155
    plasticizing effect of, 171–172, 173,
      174–175, 182
    solubility of, 13, 23, 25, 117, 123, 150,
      160, 185, 218
    sorption kinetics and diffusivity of, 15,
      185, 186, 224–225
  HFC-152a
    in foam processing, 156
    physical properties of, 155
    solubility of, 150
  HFC-245fa
    degassing of, 177
    expansion profiles with, 179–182
    in foam processing, 156, 157, 158, 160,
      168–170
    physical properties of, 155
    plasticizing effect of, 172–173
    solubility of, 177
  in foam processing, 147, 154, 156, 160
  nomenclature for, 145
  physical properties of, 146, 155
Hydrofluoroethers (HFE), 159
Hydrostatic
  compressibility, 200
  pressure, 216–217
  stress, 27

I

Immiscibility, *see* Polymer blends, immiscible
Impact resistance, 109
Impregnation, 2, 23, 125, 126
Inert gases, 122, 133, 145, 149–153, 160, 165
Injection molding, 27, 30, 49, 70, 153, 161, 226
Injection of gas into extruder, 224, 225
Insulation
  electrical, 236
  panel, 31, 147, 160; *see also* Vacuum
    insulation panel
  thermal, 31, 118, 122–123, 127, 131, 133,
    147, 154
Interface modification, 107, 111
Interfacial surface energy, 219, 223
Interfacial tension, 110, 111, 112, 113, 119, 196
Intergovernmental Panel on Climate Change
  (IPCC), 155
Interpolymers, *see* Styrene copolymers
Intrinsic viscosity, *see* Viscosity, intrinsic
Ionic interaction, 120, 132
Ionomers, 120, 126, 132, 133
Irradiation, *see* Crosslinking, irradiation
Isobutane, *see* Hydrocarbons
Isopentane, *see* Hydrocarbons
Isopropanol, *see* Alcohols

K

Kneading blocks, 224–225
Kyoto Protocol, 187

L

Langmuir
  capacity constant, 5
  contribution to solubility, 4–5
  isotherm, 4, 12
Layered foams, 187
LDPE, *see* Low-density polyethylene
Lennard-Jones
  force constant, 11
  temperature, 12
Light scattering technique, 199
Linear low-density polyethylene (LLDPE)
  energy of activation of flow of, 63, 79, 81
  gas transport (diffusivity, solubility) in, 18
  in foamed polymer blends, 85, 119, 124,
    130, 131, 133, 245
  rheological behavior of, 62, 71, 73, 80 85,
    91–92, 119
Linear viscoelasticity, *see* Viscoelastic, linear
  domain
LLDPE, *see* Linear low-density polyethylene